WEST HAM
EAST HAM
Stratford Marsh
BOW
POPLAR
Plaistow
Plaistow Marsh
Wallend
Barking Level
R. Roding
Outfall Sewer
East Ham Level
Canning Town
Beckton
R. Thames
Victoria Dock
Royal Albert Dock
King George V Dock
Silvertown
North Woolwich
W. India Dock
Isle of Dogs
Millwall
Cubitt Town
Bugsby Marshes
Greenwich Marshes
Royal Dockyard
Royal Arsenal
Plumstead
WOOLWICH
Charlton
PLUMSTEAD Common
GREENWICH
Royal Naval College
GREENWICH PARK
Woolwich Common
Shooters Hill
Kidbrooke
BLACK HEATH
Eltham Common
Welling Sta.
Oxleas Wood
Shepherdleas Wood
LEWISHAM
Lee
Lee Green
Hither Green
Catford
Rushey Green
Well Hall
Eltham Park
ELTHAM
Eltham Palace
Middle Park
Horn Park
Blackfen
Avery Hill
South End Hall
New Eltham
Halfway Street
Golf Course
Woodcroft
Eltham Coll.
Mottingham
South End
White Horse Hill

THE LONGEST NIGHT

THE LONGEST NIGHT

VOICES FROM THE LONDON BLITZ

Gavin Mortimer

WEIDENFELD & NICOLSON

First published in Great Britain in 2005
by Weidenfeld & Nicolson

1 3 5 7 9 10 8 6 4 2

A CIP catalogue record for this book
is available from the British Library.

ISBN 0 297 84638 8

Printed in Great Britain by
Clays Ltd, St Ives plc

Weidenfeld & Nicolson
The Orion Publishing Group Ltd
Orion House
5 Upper Saint Martin's Lane
London, WC2H 9EA

www.orionbooks.co.uk

Contents

Acknowledgements

Scattered around the world is a small band of Londoners unwilling or unable to rid themselves of the memories of the Blitz of 1940–41. Despite the sixty or so years in which they've grown old and bent and weary the Blitz is still there, in the dusty cellars of their minds. Sometimes it emerges. It might be a car backfiring or smoke rising from a burning building. Then the memories return.

When I appealed in various magazines and local papers for the reminiscences of Blitz survivors of May 1941 I was utterly unprepared for the response. Nearly two hundred letters dropped on to my doorstep, written with eloquence, humour, modesty, honesty and clarity: from Canada, New Zealand, Australia, South Africa, USA, Gibraltar, Spain and all over the British Isles. A collective thank you for reliving old and sometimes painful memories for the benefit of a complete stranger, and particularly to the dozen or so correspondents who were happy to expand by phone on various points that arose from their letters: Gwen McWilliams, a wartime ambulance auxiliary; David McCarthy; Nick Allen; Edna West; Stanley Lathwell (who lived next door to, not *in* Wormwood Scrubs Prison); Eileen Bain; Morag Storie; Eileen Martin and Iris Strange.

In addition, I interviewed in person several Blitz survivors who, though they don't feature in the book, furnished me with invaluable insights into life in London during the period: Frank Marshall; ex-firewomen Joan Jones and Mitzy Spooner; Charles Harwood and Cyril Demarne, two wartime firemen in their late 90s; Patrick Kinna, Winston Churchill's short-hand clerk during the war and a man with a hatful of amusing anecdotes; Leslie Scott, an AA gunner (who has since died); Morrie Foley, an East End docker; Dave Clark; George Pledger; Fred and Joyce Newman; Patricia Whenham; Innes Michell, one of the last remaining WVS volunteers from the Blitz; Dorothy Parker; Pat Mansfield; Amy Hughes; and Charlie Dench, a teenage recruit in the East Ham Home Guard.

Apart from first-hand accounts, *The Longest Night* draws on numerous secondary sources. Thanks to the following for all their assistance and generosity: Joan Boucher, editor of the Doddington Parish newsletter, who put me in touch with the Veazey family. And Zoe Huckstepp and her granddad, Peter.

The Royal Air Force was its normal efficient self in unearthing information. Bob Cossey, secretary of 74 Squadron Association, warrants a special mention for allowing me to publish the photo of Johnny Freeborn and Roger Boulding, and I recommend his book *Tigers: Story of 74 Squadron.* Geoff Faulkner, of 264 Squadron, filled in the gaps in Freddie Sutton's service record; and the family of the late Horace 'Danny' Kilner were kind enough to grant me permission to dip into Sutton's memoirs. Wing Commander Jeff Jefford of the RAF Historical Society, Derek Rothery of 85 Squadron and the staff at the RAF Museum Hendon were also of great service.

The Army yielded nothing to the RAF in efficiency. Thank you, Captain Wilson and James Armstrong of the Honourable Artillery Company, Rebecca Cheney, curator of the Royal Engineers Museum, and everyone at the Bomb Disposal branch of the Royal Engineers Association.

Barry Sergeant of the Retired Firefighters Association invited me to a wartime reunion of London firemen and women – an unforgettable day – and Esther Mann, curator/archivist of the London Fire Brigade Museum, was unceasingly cooperative, as was Terry Jones, of the LFB photo archive, who rooted out some of the stunning photos within these pages. Thanks, too, to Elaine Willis for sourcing other photos and to Jo Murray for proof-reading.

I take off my hat to Helen Kent, in charge of the London Transport Museum archives, Susan Snell of the Transport for London archives, Linda Robson, curator of the Barking and Dagenham Museum/Archives and David Williams at Westminster Library. I haven't names for the people who pointed me in the right direction at the following but you have my sincere gratitude: the borough archives of Lambeth, Tower Hamlets, Croydon, Islington, Camden, Southwark, Hackney, Newham and Hammersmith; the Metropolitan Archives in Clerkenwell, the staff at Lambeth Palace Library, the library staff at London Fire Brigade HQ on the Albert Embankment, the National Newspaper Library and all those at the Imperial War Museum, not just for aiding my research but for allowing me to quote from Ballard Berkeley's interview and from the papers of Vera Reid and John Hughes.

Thank you to Nadine James of Southwark Council and Rose Lingham and Ian Adams of the Southwark Pensioners' Association for putting me in contact with so many Blitz survivors, and likewise to Bill Dempsey of the East Ham Pensioners' Association and Angela Sinclair in Islington.

Albert Hufenreuter junior answered all my questions concerning his father with patience and honesty, and Richard Collier's book, *The*

City that Wouldn't Die (Collins, 1959) furnished me with some additional information concerning the Luftwaffe. Kristin Ecuba, Gernot Bauer and Damon Allen helped with translating German documents. The following did useful groundwork on various Blitz-related themes: Graham Church, Roy Carr, Jeffrey Palmer, Arthur Thomson, Anne Harwood, Dot Gibson, Hugh Ross, David Harkins, *Woman's Weekly*, *The British Pensioner Magazine* and Beryl Furey-King.

I have tried to contact all copyright holders of material quoted. Where I failed, I apologise. If the people concerned would contact me I will happily acknowledge their generosity in future editions of *The Longest Night.* Meanwhile, my thanks for their kind permission to quote from published material to: the trustees of the RAF Museum Hendon for Guy Gibson's *Enemy Coast Ahead*; the trustees of Lambeth Palace Library for Alan Don's diary; David & Charles Publishers for Neil Wallington's *Firemen at War*; the trustees of the London Transport Museum for Charles Grave's *London Transport at War*; the Corporation of London, London Metropolitan Archives for Anthony Heap's diary ACC/2243; Barking & Dagenham Local Studies Centre at Valence House for the war memoirs of T. A. Cockburn; Eric Bathurst for the diary of Gertrude Bathurst; Secker & Warburg for Sefton Delmer's *The Black Boomerang*; and J. M. Dent for Madeleine Henrey's *London Under Fire*.

Thank you to my agent, Andrew Lownie, my editor Penny Gardiner, who took time off from wiring houses to do a first-rate job as ever in spotting where I was labouring a point or simply waffling, and to Ian Drury at Weidenfeld for the unobtrusive way in which he lets me get on with it.

The Longest Night was researched and written in 2004, an arduous challenge completed only with the unstinting encouragement, love and patience of my wife, Sandy. That she could provide such sterling support while pregnant is testament to her French fortitude! Two weeks after I finished writing *The Longest Night*, Sandy gave birth to our first child, Margot. My love for you both has no limits.

For many, London is an ugly, dirty, sprawling mass of a city. They're right, of course; but the greatness of London has always lain not with the beauty of its buildings or the elegance of avenues but in its people, who are cussed and defiant, funny and warm. This book is a tribute to those nameless civilians who remained true to their city when it was wounded and hurting and in need of their loyalty. In particular it is a salute to Tom Winter, John Fowler, Joan Veazey, Gladys Jenner, Gladys Shaw, Joan Wallstab, Joe Richardson and Emily Macfarlane, who took it, every last bomb of the Blitz, with a quiet courage and a dauntless spirit.

Gavin Mortimer

MONTPELLIER 2005

Introduction

'SIEG HEIL!' 'Sieg Heil!' 'Sieg Heil!' Hitler stood before his acolytes, smirking contentedly at their chants of adulation. Another rally, another taunt at the British. Here in Berlin he had a special message for those defiant little Islanders. He surveyed the throng of nurses and social workers gathered before him and gestured for silence. 'In England they're filled with curiosity and keep asking, "Why doesn't he come?"' The crowd murmured as Hitler paused. 'Be calm, be calm,' he continued. 'He's coming! He's coming! When the British Air Force drops two or three thousand kilograms of bombs, then we will in one night drop 150, 250, 300 or 400 thousand kilograms. When they declare that they will increase their attacks on our cities, then we will raze their cities.'

Three days later Hitler fulfilled his promise. Three hundred and fifty German bombers, escorted by 600 fighters, blotted out the warm September sun as they came up the Thames Estuary. They were 2 miles high and the head of the force stretched across 20 miles. At 5 p.m. the bombs began to fall. On Woolwich Arsenal, the country's largest munitions factory, and then on the docks that proliferated along the banks of the Thames. But this was a day not just for military and industrial targets, this was a special day. Saturday 7 September 1940 was the day Hitler had chosen to launch his attack on the British people. This was the beginning of the Blitz. Bombs spewed out of aircraft and fell with murderous intensity on London's East End. People in the boroughs of East Ham and West Ham, in Stepney, in Poplar and in Whitechapel died easily in the late summer sunshine. There were 430 dead as night edged in, and elsewhere across London people watched with bemusement: why, they asked each other, was the sun setting in the east as well as the west? The fires guided in the second wave of German bombers at 8 p.m. When dawn broke on Sunday 8 September, 400 more were dead.

THE British had been expecting the Blitz, the 'Knockout Blow', for a long time. Way back in 1937 an Air Raid Wardens' service had been formed. The following year the Government distributed a pamphlet asking for another 30,000 people to join the 150,000 volunteers already in the Air Raid Precaution (ARP). They were after 'Men over 30 and women over 18 who by reason of their experience or occupation or their position in social or public life, possess standing or influence in the immediate neighbourhood of their homes. Telephone desirable but not essential.'

A few months later the Civil Defence was mobilized as Hitler demanded Czechoslovakia. In the days before Chamberlain appeased the Nazis, the spectre of war had galvanized half a million more Britons into volunteering for the Civil Defence. The newcomers were put to work at once, digging trench shelters in parks, building brick street shelters and sandbagging important buildings. At the same time barrage balloons, silvery like sardines, were winched up over London, and 38 million gas masks were handed out. (A gas attack was what the people feared most, having learned of its horrors from veterans of the Western Front.)

Eleven months later, on 24 August 1939, the Civil Defence were primed for action as Germany eyed up Poland. On Friday 1 September they invaded and Britain's Civil Defence, 1.5 million men and women strong, was once more mobilized. As well as Air Raid Wardens, there were auxiliary firemen and women, auxiliary ambulance drivers and reserve policemen. Middle-aged women joined the Women's Voluntary Service (WVS) and helped in the evacuation of children and the provision of food to the Civil Defence.

As London's children were evacuated a handbook was sent to each family by the Ministry of Home Security entitled *Air Raids – what you must know, what you must do*. The information provided was basic and banal and it most certainly did not include the Air Staff's conclusions about the capability of the German Luftwaffe, which, in light of their performance in the Spanish Civil War, had surmised that 800 German aircraft were capable of dropping 950 tons of high explosives on Britain per day. It therefore seemed likely, the Air Staff concluded, that in the first six months of a war 600,000 Britons would be killed in air raids and another 1,200,000 wounded. Plans were discussed about the best way of disposing of the dead, proposals including mass burials in lime pits and the dumping of bodies in the sea. On the forecourt of Victoria Station a tower with a loudspeaker had been constructed so that when the expected hordes of panicked Londoners descended with the intention of fleeing to the country they could be controlled with greater efficiency.

But the German planes didn't show up. When the air raid sirens over London sounded on Sunday 3 September, just hours after Britain's declaration of war on Germany, it turned out to be a false alarm; nothing more than a disorientated French aircraft. Instead the 'Phoney War' began. To some it came to be known as the 'Sitzkrieg', for others it was the 'Bore War'. It wasn't long before the Civil Defence, particularly the 400,000 full-timers, were being derided in the Press. They were busybodies whose only purpose seemed to be enforcing the blackout every night. At first people hurled abuse at the wardens in the street, then words were replaced by boots and fists. Responding to the criticism, the Government pared their numbers and by the end of 1939 there were only 200,000 full-time personnel in the Civil Defence.

The arrival of 1940 brought nothing new. The London-based journalist, Molly Panter-Downes, wrote in the *New Yorker* that it had been a 'war of yawns' for the capital's Civil Defence. 'They spend their nights playing cards, taking cat naps and practically yearning for a short air raid.' By late spring of that year most evacuees had returned to London, but the ravaging of the Low Countries by Germany in May sparked a fresh exodus. In six days 180 trains left the capital carrying 100,000 children out of the putative danger zone. Holland and Belgium fell, then France. The British Expeditionary Force was plucked from Dunkirk and the country girded its loins for invasion. But the battle for Britain was fought in the air, not on the beaches.

On 30 June, Colonel General Jodl, Chief of the Armed Forces Command Staff, issued a memorandum called 'The Continuation of War against England'.

> If political methods should fail to achieve their object, England's will to resist must be broken by force:
>
> a) by attacks upon the English homeland
>
> b) by an extension of the war peripherally.
>
> So far as (a) is concerned, there are three possibilities:
>
> (1) Siege. This includes attack by land and sea against all incoming and outgoing traffic. Attack on the English air arm and on the country's war economy and its sources as a whole.
>
> (2) Terror attacks against the English centres of population.
>
> (3) Invasions with the purpose of occupying England.

> The final victory of Germany over England is now only a question of time. Offensive enemy operations on a large scale are no longer a possibility.

But throughout high summer Hermann Goering's Luftwaffe tried to pummel the RAF into submission. On 15 August, the day known as *Adlertag*, the Germans launched their biggest attack yet, but it was a failure. Seventy-five aircraft were shot down (not the 182 claimed by the British). They reappeared three days later but were repelled once more. Thick cloud prevented further assaults until 24 August. Portsmouth was bombed. One hundred died, the heaviest civilian casualties suffered by Britain so far. On the same evening a horde of German bombers took off from France to attack oil refineries and factories in southern England. Several aircraft lost their way over the blacked-out landscape and dropped their bombs on central London. Churchill demanded an eye for an eye and the next night forty-three RAF bombers hit Berlin. The raid achieved little, other than to incur the wrath of Hitler. For months he had ordered that London be left alone until the day before the invasion of England began; then the Luftwaffe was to submerge the capital in bombs and 'cause the population to flee from the city and block the roads'. Churchill's impertinence in attacking Germany infuriated Hitler. For the next few days and nights the RAF continued their assaults which, although largely ineffectual, were hugely embarrassing for the Fuehrer who had promised his people immunity from attack.

On 28 August RAF bombs killed a number of Berliners. Hitler's reprisals came quickly, with four innocuous raids on Liverpool on the last nights of August. But at the start of September he oversaw a fundamental shift in German strategy.

Since 24 August the Luftwaffe had concentrated its fury on the RAF's Fighter Command airfields. Six of the seven sector stations had been bombed to the point of oblivion. After weeks of heroic resistance the RAF was on its knees. On 4 September, the day Hitler addressed the rally of nurses and social workers in Berlin, the RAF lost nearly three hundred aircraft and one hundred of its pilots. Factories couldn't produce enough new planes and the replacement pilots were horribly raw. If the onslaught continued for another week, the Battle of Britain would go the way of Germany. Faced with such a parlous situation Air Chief Marshal Dowding ordered his pilots – hitherto more belligerent than their German adversaries – to avoid contact if at all possible.

The Luftwaffe knew they were hurting the RAF but they had no idea the pain was so great; rather they were growing ever more alarmed at their own casualties. In two weeks a hundred of their twin-engine bombers and two hundred fighters had been shot down. Scores more had limped back to their base with extensive damage that necessitated lengthy repairs.

A blitz on London would meet both Hitler's and the Luftwaffe's requirements. Hitler would have his retribution and the Luftwaffe would have a glorious opportunity to finish off the inferior numbers of the RAF once and for all as they scrambled to meet the threat.

The final consideration was the weather. The perennial obsession of the British had for the last few weeks obsessed the German High Command. The navy had informed them that 21 September was the last day in that month when conditions would permit a seaborne invasion. Failing that, the weather could not be relied upon until the spring of 1941. As it was judged that any invasion fleet would need ten days from the day of the decision to invade to actual embarkation, 11 September was the cut-off date for the mobilization of Operation Sealion. On 6 September Hitler gave permission for the first major air raid on London. Twenty-four hours later 950 Luftwaffe aircraft took off from their French airfields to bomb London. As they crossed the

Channel on Saturday afternoon, 2 miles up in the sky and covering 800 square miles, Goering waved them on their way from the cliff tops of Cap Blanc Nez.

BY 11 September, London's population had been sheared by the Luftwaffe. Hundreds were dead, but it was defiance not defeat that infused their blood. In Stepney somebody stepped over the corpses and the rubble and on a wall scrawled a message to Hitler: 'England for ever. Keep smiling. He may get us up. But he'll never get us down.'

Hitler decided to give his air force more time to annihilate the RAF and the deadline of 11 September was extended. Churchill braced his people for another onslaught by appealing to the Briton's traditional love of the underdog. The next week, he told them, would be a momentous one in British history, worthy of comparison 'with the days when the Spanish Armada was approaching the Channel, and Drake was finishing his game of bowls; or when Nelson stood between us and Napoleon's Grand Army at Boulogne'.

The decisive day was 15 September. More than fifty German aircraft were shot down on their way to bomb London; the numbers weren't particularly significant, but the morale of the survivors was damaged. They had been assured that the RAF was terminally ill, but it was clear that far from being in their death throes the British pilots were in rude health. Two days later Hitler postponed the invasion.

The heavy daylight raids continued for the rest of the month. Crucially, the Germans attacked not just the poor East End of London but the more affluent central and western boroughs. If they had stuck to the working-class heartlands who knows what reaction they might have provoked in areas already seething with social injustice. 'If only the Germans had had the sense not to bomb west of London Bridge,' Clement Attlee told Harold Nicolson, 'there might have been a revolution in this country. As it is, they have smashed about Bond St and Park Lane and readjusted the balance.'

In early October the Luftwaffe stopped daylight raids and began bombing exclusively at night. Londoners started to lead two separate lives; the quite routine of 9 to 5 and the terror that came with the dusk. Any possible invasion was out of the question for at least six months, but the raids continued because Hitler wanted to spread defeatism and demoralization in London, and because he enjoyed killing people.

LONDON'S first respite since 7 September was 3 November, when the weather was too filthy even for the Luftwaffe. Little wonder that the US Ambassador had notified his government that in his opinion Britain wouldn't survive much longer. Around this moment, the American magazine *Time* ran an article entitled 'The Legacy of Britain'. Londoners scoffed at the headline; so did Churchill. He had never any doubt that the capital would survive. 'I was glad that, if any of our cities were to be attacked, the brunt should fall on London,' he wrote in his war memoirs eight years later. 'London was like some huge prehistoric animal, capable of enduring terrible injuries, mangled and bleeding from many wounds, and yet preserving its life and movement.'

On 14 November the Luftwaffe took the Blitz to the rest of the country. Coventry was first but over the next few months most large ports and cities in Britain were targeted, or 'Coventrated' as the Germans jeered over the airwaves. Some of the ordnance being dropped on their heads, Lord Haw-Haw gloated, had been left behind in France by the RAF six months earlier. The Luftwaffe was 'returning it'.

London was hit hard again on 8 December. A Christmas truce, brokered by the German Embassy in Washington, ended on 29 December when the City was firebombed and 163 killed. That took London's death toll since 7 September to 13,339, with 51,693 wounded.

Death had long since lost its sting at the start of 1941. Londoners had become inured to most things, be it a baby sucked out of a house by the blast of a bomb, its brain dashed out on the road, or the Air Raid Wardens who scoured bomb sites collecting body parts in sacks for possible identification at the mortuary.

THERE was another heavy raid on 11 January, then a little era of peace. It ended on 8 March, when 125 aircraft attacked London, ninety-four aircraft decorating the city with their bombs on the following day. They were back on 15 and 19 March. A lull for a month, giving time for the daffodils to appear and the cheering smell of spring to lift the city's spirits. Then on 16 April the largest force of bombers yet seen, 685, laid waste to London in what came to be remembered as 'the Wednesday' raid. One thousand were killed. Worse was to come three days later, 'the Saturday', when 712 planes mounted a reprisal raid for an RAF attack on Berlin. The number of dead exceeded Wednesday's tally by 200. Never had number crunching been so grim. The 1,200 added to the existing total brought London's fatalities since September to more than 16,000, with another 23,000 seriously wounded. Not a patch on the original pre-war forecasts of the Air Staff but severe enough to shake the capital's morale. Continual bombing, day after day, night after night, had been easier to deal with psychologically. People knew what to expect, and when to expect it. The first fifty-seven uninterrupted days of bombing had come nowhere near weakening people's resolve. 'We can take it,' they boasted. But the raids of March and April had been different: they'd been heavier, far heavier than anything in 1940, and the 'Blitz Nerves' of many were fraying. It was the suspense that excoriated them. Will they come tonight? Maybe tomorrow? Perhaps next week? Or when? They didn't know. All they knew was that they would come again and when they did it would be in force.

An article in the *New Statesman* encapsulated the mood of many: 'Life never seemed so unreal, so like a chapter from a novelette, as it has done this week,' wrote the journalist. 'When I get up in the morning I have actually to look at the damage before I can believe that so many of the buildings and places that I most treasured in London have just disappeared off the face of the earth.'

As May replaced April London had little need to worry. Hitler had grown bored of the Blitz and was anyway now after a bigger prize than insignificant Britain. His lust was for Russia and several bomber units had already been withdrawn east in readiness for the imminent invasion. The culmination of the eight-month ransacking of Britain was a week of violence against Liverpool, Glasgow, Sheffield and Hull. After that, only a token force of Luftwaffe bombers would remain in France.

But on Thursday 8 May the RAF retaliated. Nearly four hundred bombers attacked the cities of Bremen and Hamburg, inflicting the heaviest damage of the war so far on Germany, though the ninety-four people killed in Hamburg was a paltry number by comparison to the casualties suffered in dozens of British cities in the past nine months. Nevertheless, Hitler was incensed. Such effrontery must not go unpunished. The twenty or so guests who gathered at his Bavarian chalet retreat 8,000ft above the town of Berchtesgaden on Friday 9 May were there to hear him hold sway on a range of issues. As they stood in the gold-leafed elevator that took them from the car park into the chalet they suspected the raids would be mentioned. It was an esteemed coterie that were led into the Great Room, among them Martin Bormann, the Nazi Party Chairman, Feldmarschall Wilhelm Keitel, Commander of the German Armed Forces, and Hans Baur, Hitler's personal pilot.

It was still daylight and the views north from the grand picture window over the Bavarian Alps were awesome. For a few moments they stood in reverential silence. In the corner to the left of the window was a large revolving globe, smudged with the fingerprints of Hitler as he dreamed of further domination. Around the walls were Persian carpets, Gobeline tapestries, Rembrandt paintings, all of them looted by the Nazis from their conquered lands. As Eva Braun greeted the guests the room was warmed by the flames from a marble fireplace in the south wall. Above the mantelpiece was a Botticelli nude. It looked on impassively as Hitler gave vent to his fury against the British.

IN London it was an unremarkable Friday evening. Double Summer Time had been introduced on 4 May and the winding forward of the clocks by two hours was like a sweet tune of hope. Daylight lingered until after 10 p.m., and across London people took advantage of the long evenings. Colder than normal for this time of year, perhaps, but that wasn't the only reason for the red glow in people's cheeks. It was three weeks now since the last heavy raid on the capital. Regular sleeping patterns were being woven and the tension that had gnawed at their bones was beginning to disappear.

The newspapers encouraged optimism, detailing the dominance of the RAF nightfighters over the Luftwaffe bombers. On Thursday the *Evening Standard* had bragged that the 'Nazis lose 290 airmen in a week'. Friday's edition of the *News Chronicle* printed a photo of the King inspecting a squadron of RAF nightfighters under a headline 'Nazis lose 38 raiders in 20 hours'. A new office pastime had sprung up: guess the number of German bombers shot down each night: 3d a guess with 10 per cent going to the Red Cross, the rest into the winner's pocket.

In Croydon, on the outskirts of south London, householders scrunched up Friday's *Croydon Advertiser* and used it to start their fires. Few would have spotted the small article tucked away inside, lost among the wills and small adverts. It was entitled 'Horoscope' and for the unbelieving and the sceptical it was hilarious claptrap: 'It is well over a year ago that a speaker at a religious meeting in Croydon gave May 11th 1941 as a critical day in the experience of this nation,' began the article. 'His forecast was based on prophecy derived from an interpretation of measurements in the interior of the Great Pyramid according to which May 11th represented either the entrance or exit of the Second Low Passage that leads into the terminal space known as the King's Chamber.' Readers guffawed and read on to the final passage. 'Whatever May 11th or any other day may bring, let us be found steadfast in adversity and modest in triumph.'

1

FELDMARSCHALL Hugo Sperrle was not a handsome man. His hairline had receded to the top of his crown and what was left of his dark hair was slicked back with handfuls of oil. His eyes were permanently narrowed in hostile suspicion. Even when he was whoring in the best brothels of Paris or gambling in the casinos of Deauville, Sperrle's eyes were lifeless. But his ugliest feature were his jowls, repulsive folds of flab that draped down over his jawline, bringing the corners of his mouth with them. The result was a permanently baleful expression at odds with his well-fed physique.

But today Sperrle's spirits had received a sudden and unexpected boost, despite the ungodly hour at which he had been awoken in his sumptuous quarters at the Palais du Luxembourg in Paris. The call had come from General Hans Jeschonnek, Chief of Staff to the Luftwaffe, from his office in Hitler's Wolfsschanze HQ (Wolf's Lair) in East Prussia. It was shortly after 8 o'clock on the morning of 10 May 1941, but Jeschonnek didn't apologize. Hitler had been on the phone from his summer residence in Berchtesgaden. He wanted London bombed, tonight. The RAF attacks on Bremen and Hamburg must not be allowed to go unanswered. Jeschonnek passed the message to Sperrle, reminding him that the bombers of Feldmarschall Kesselring's Air Fleet Two were at his disposal as Kesselring was in Poland.

The telephone call was brief. Sperrle wasn't a man who used two words where one would do, and Jeschonnek knew that he would plan and execute the raid with ruthless efficiency.

As Sperrle clothed his massive girth in his grey air force uniform, he glanced out of the window of his quarters in the Palais, built in the Florentine style for Marie de Medici in the early seventeenth century. The sun had just breasted the trees on the eastern side of the Jardin du Luxembourg. The octagonal pond, the *Grand Bassin*, beneath his window was dappled with sunlight. He secured his monocle over his right eye and sneered as he thought of the fate that awaited London.

Even though his Chief of Staff, General Karl Koller, was waiting for him in his office, he didn't rush. Dressing was never straightforward for Sperrle. There was that much of him to cover, and his fondness for livery was as extravagant as that of Hermann Goering, his Luftwaffe commander. 'His craving for luxury and public display ran a close second to that of his superior,' wrote Albert Speer, after a visit to the Palais du Luxembourg, adding: 'He was also his match in corpulence.'

When dressed to his satisfaction, Sperrle joined Koller in front of a large and detailed map of London. Sperrle grunted, mystified by Hitler's order. It showed typical inconsistency. One moment they were instructed to husband their aircraft and crews because of the impending invasion of Russia and the next they were to despatch every available aircraft to annihilate London.

Despite his oafish appearance Sperrle was no fool. The snobbish disdain of his peers was motivated by envy. How could the son of a Wurttemberg brewer rise to such heights?, they mulled. They ignored the brilliant strategic mind that had been sharpened by his experiences as a pilot in the First World War. Subsequently, Sperrle had risen through the ranks until he was given the command of the Condor Legion in the Spanish Civil War, contributing to the development of saturation bombing through the devastation of Guernica, Aragon and Ebro. He had argued against Goering's decision the previous September to switch from attacking British Fighter Command airfields to launching huge air raids on British cities and strategic targets. His flabby jowls had wobbled indignantly as he argued his case with Goering; the British Air Force must be first obliterated before British cities are targeted. It would be folly to do otherwise and Sperrle suspected the RAF wasn't quite as feeble as Goering believed.

But Sperrle's concerns were brushed aside. Piqued by Goering's indifference to his views, Sperrle took out his anger on the British population by evolving the bombing techniques first practised in Spain in the 1930s.

The first attacks on London in September had followed the same pattern as those in Spain: blanket bombing concentrated on specific areas of the city. The results had been pleasing, but later in the year he began to adopt a different method: hundreds of small incendiary bombs were dropped to start myriad fires and, as the emergency services fought the fires, the second wave of bombers delivered hundreds of high explosive bombs. Mayhem was what Sperrle craved, and at first he was delighted with the reports he received. But he soon realized that high explosives dropped late in a raid often came as an unlikely salvation to the firemen, creating firebreaks that prevented flames from engulfing great swathes of London. So Sperrle evolved a new and callous technique that played on the city's fickle volunteer firewatchers. High explosive bombs were dropped first, to scare the firewatchers back to their homes, and then incendiaries followed, falling on the unguarded offices and warehouses.

Sperrle squinted through his monocle at the map in front of him, as the sun continued its rise above the Jardin du Luxembourg. He had five hundred aircraft to send across the Channel, but what should they target? Decisions would have to be made quickly. He stabbed his fat forefinger at the Thames Estuary, just beneath Canvey Island, and moved it upstream from Canvey, tracing the route the Luftwaffe navigators used when they flew in over Sheerness and picked up the Thames. Past Gravesend and Tilbury, up to Dagenham, battered in October 1940 with many fatalities, on to Barking, the recipient of fifty-four high explosives in the last heavy raid of 19 April. Sperrle's finger stopped momentarily at Barking Creek, which ran into the Thames by the Beckton Sewage works. He smiled, recalling the British embarrassment when, on 6 September 1939, two Spitfire pilots had mistaken a pair of Hurricanes for enemy aircraft and shot them down in what became known as the Battle of Barking Creek. It had caused much amusement within the Luftwaffe.

Sperrle's finger sailed on, following the Thames as it curved south after Barking, past Thamesmead and into the stretch of the river known as Woolwich Reach. Woolwich Arsenal, Britain's biggest munitions factory, lay on the southern bank, the Victoria Docks on the northern shore. Both had been pulverized in the September raids.

Sperrle's finger skimmed across the Thames and followed its course as it curved past West Ham and round the tip of North Greenwich, shaped like a hitchhiker's thumb. He paused, his finger hovering at the top of the river's U-bend. His pilots cherished this bend, the landmark they used to guide them on the final approach to their target. Here the river arced south towards the National Maritime Museum and Greenwich Park, where it began to coil north. In the middle of the bend was the Isle of Dogs, with the Millwall and West India docks – where grain and rum and sugar arrived from around the Empire – at its noisy, vibrant heart. Across the river were the Surrey Commercial Docks, where the bombing of 7 September had ignited vast piles of Norwegian softwood timber stored within its 250 acres.

Now Sperrle's finger journeyed into tourist waters. Up ahead was

Tower Bridge, with the Tower of London in its shadow on the north shore. A mile further west was London Bridge. Now the bridges came thick and fast: Cannon Street railway bridge, Southwark, the two Blackfriars bridges, only 50 metres apart, with a battery of AA guns on railway wagons at either end of Blackfriars railway bridge to protect the ammunition trains that used the line.

Here Sperrle's hectoring finger jumped out of the water and became a commuter, taking the path thousands of south Londoners trod every day on their way to work in the offices of London's financial district. North down New Bridge Street, past the turning on the right into Queen Victoria Street, to Ludgate Circus. Left was Fleet Street, from where the printing presses of Britain's newspapers ran red hot with scorn for Germany and its leaders, and the church of St Clement Danes at the western end. Right was Ludgate Hill, which led past the Old Bailey and to St Paul's Cathedral. Sperrle shook his head. St Paul's Cathedral. How had this treasured icon of the British survived for so long, he wondered. He thought of 12 September, when a 1,000kg bomb had hit the cathedral but failed to explode. Perhaps this time?

Sperrle's finger headed back over Blackfriars Bridge like one of the legions of commuters returning home. Now he was in Southwark, going straight down the Blackfriars Road, over and across St George's Circus, into London Road for a quarter of a mile and there was the Elephant & Castle. He tapped the map lightly a couple of times. The Elephant & Castle, what an absurd name. Its junction resembled a six-armed starfish, each a vital thoroughfare. Sperrle deliberately followed each arm of the starfish in turn: Newington Butts led south-west to Kennington and Brixton; the Walworth Road went south to Camberwell and Peckham; the New Kent Road ran east to Tower Bridge; Newington Causeway took you towards London Bridge; London Road, he knew, was Blackfriars, and St George's Road ...? Sperrle's finger ran north-west, along St George's Road, into

Westminster Bridge Road, skirted the southern end of Waterloo Station and there his finger was faced with a choice: left along Lambeth Palace Road to the Tudor brick gatehouse behind which the Archbishop of Canterbury lived, or straight over Westminster Bridge to Big Ben, Westminster Abbey and the Houses of Parliament. Wouldn't it be satisfying if we could rid the world of Herr Churchill?

Sperrle's finger finished its journey across London and began stroking its master's bulging double chin. By 9 a.m. he and Koller had divided the English capital into three generous portions. The East End and the City's financial district; Westminster and north as far as King's Cross Station; and south of the river from Greenwich to Clapham in the west. In the centre of this third target zone lay the Elephant & Castle.

2

EMILY Macfarlane waved goodbye to her mother and walked out of the Peabody Estate. It was 9 o'clock on Saturday morning and London was waking with a cheery optimism. Emily turned into Rodney Road as the sun burned gold in an all-blue sky. It was shivery cold, though, and with every breath a plume of steam dashed out in front of Emily's face. She reached up with her gloves and pulled her beret tight, so that her ears were covered.

She walked north along Rodney Road with a staccato of high heels on the frosty pavement and passed her old school on Victory Place. She thought of her three former classmates killed in the Blitz. Past the school, Rodney Road became Rodney Place, both streets named after Admiral George Rodney, who in his time had despatched countless French and Spanish ships to the bottom of the sea.

Now Emily was walking west down the New Kent Road. In five minutes she'd be at the Elephant & Castle. She drew admiring glances from the men who passed her on the pavement, most of whom had to look up to catch a glimpse of her pretty, catch-me-if-you-can face. Born into a poor south London family she may have been, but Emily had an elegance and class way above her station. She was tall, 5ft 10ins, with brown hair that hung down from her beret to below her shoulders. Though she had plenty of suitors, Emily was single, despite the temptations that marriage offered. Not the temptation of love and romance, but marriage was the only route out of the fire service into which she had been conscripted. Her great friend, Joan Ellis, had wed at the end of 1940 and she was now exempted from conscription.

For Emily the fire service was the bane of her life; she hated the uniform, the unfashionable cut of the skirt and the fact she had to wear fawn stockings with clumpy black shoes. She loathed the fireman whose wandering hands had earned him a stinging red handprint of Emily's on his face; and she was bored with sitting in the control room at her fire station taking messages about falling bombs. She was back on duty later this evening at the sub-station just off the New Kent Road for the start of a 48-hour shift. But now it was time to shop at the Elephant. Perhaps they might have some stockings in? A thought flashed into her mind: why not call on Joan to see if she wanted to come? But then she remembered that Joan was down in the West Country, visiting her husband at his army base.

THE route Emily was taking towards the Elephant & Castle had at one time been walked by some of England's greatest minds: the poet Robert Browning was born in Camberwell; the writer and social philosopher, John Ruskin, was a Southwark man; and the pioneer of feminism, Mary Wollstonecraft, wrote *Vindication of the Rights of Women* while living in Southwark. Living in the area, they had witnessed its transformation from rural idyll into one of the heartlands of the Industrial Revolution. Between 1801 and 1881 the population of Walworth soared from 15,000 to 108,000. The rich migrated further out of the city and the poor from across the river, whose homes had been pulled down to make way for factories and offices, flooded into the district. Four or five families moved into the elegant Georgian townhouses previously occupied by a single wealthy family and shops replaced front gardens on the main roads. Workshops proliferated in the backyards, making everything from fireworks to flea powder. For those too old or too sick to work there was the Newington Workhouse, a dismal repository for over 1,000 paupers.

IN 1861 the Metropolitan Tabernacle held its first service at the Elephant & Castle. The preaching of Charles Spurgeon had such a hold on the congregation that it soon became known locally as Spurgeon's Tabernacle. In those days it held 6,000, with enough seats for 3,500. For the less reverential the opening in 1872 of the Elephant & Castle Theatre at the top of the New Kent Road was a cause for greater celebration.

The other favourite theatre among locals had been the Surrey Music Hall, up the London Road by St George's Circus. One day in the 1890s a young boy called Charlie Chaplin had sat in the gallery of the Hall watching Fred Karno's *Early Birds* having scrounged the entrance fee from a local woodchopper. The Surrey Music Hall had closed down in the 1930s but the South London Palais, also on London Road, was going strong.

With the coming of the cinema at the turn of the twentieth century, picture houses sprouted up all along the Old Kent Road. The arrival of the 'talkies' was even more of an event and in December 1930 the Trocadero opened its door for the very first time. It, too, was on the New Kent Road, opposite the Elephant & Castle Theatre (which had been converted into a cinema in 1928).

THE dirt and dynamism of the Elephant & Castle attracted one of London's most famous Bohemian painters to the area in the 1930s. Austin Osman Spare had grown up in Kennington but moved to the more stylish surrounds of West London when his work found favour with London's art world in the *beau monde* of the early 1900s. One of his paintings had been exhibited in the 1904 summer exhibition of the Royal Academy when he was only 16. The critics lauded his style, describing it as having 'the hand of Dürer and the eye of Dante'.

Then in the 1920s he turned his back on his wealth and returned to his roots, renting a small and dingy studio flat in the Walworth Road. He lived sparsely with half a dozen cats for company, drawing portraits of the people he found in the Elephant & Castle and selling them in the pubs for never more than a fiver. To the locals he was the artist who looked like a tramp in his old army shirt and tattered jacket; mad as a hatter, they told each other, but you know what he'd said to Hitler, don't you? When the Fuehrer had invited Spare to Berlin in 1936 to paint his portrait, the painter turned him down. 'Only from negotiations can I wholesomely conceive you,' he told Hitler. 'For I know of no courage sufficient to stomach your aspirations and ultimatums. If you are superman, let me be for ever animal.'

Bustling in and around the shops of the Elephant & Castle were an endless stream of life models for Spare; Burton's the tailors and their fifty shilling suits; Dean's Rag Shop, purveyors of linen children's books; Colliers, number one for ladies' wear with a dancing school above; Isaac Walton, a large boys' outfitters; and a host of sweetshops, cheap jewellers, fruiters and the popular tobacconist with its own library where books could be borrowed for a penny. And there were the pubs, chock full of working men quenching their thirst, with the most celebrated of all, the Elephant & Castle pub, on an island in the middle of the junction between Newington Butts and Walworth Road, surrounded by Drapers' Almshouses. Over the pub was its symbol, a large red elephant with a castle on its back. This was the Elephant & Castle which Emily Macfarlane was entering at 9.30 on the morning of 10 May 1941.

ALSO heading for the Elephant at around the same time was 16-year-old Joe Richardson, a handsome teenager with street wisdom way beyond his years. His piercing dark blue eyes had seen more than most boys his age. He lived with his aunt and uncle in Bowyer Street, not far from Camberwell Green, in a two-up, two-down house.

Joe had left school at 13 to work as a messenger for a firm of Camberwell printers. One day in September 1940, during a daylight air raid, he'd just stepped out of the office on his way to deliver a message when he 'heard a rush of air, like an express train coming towards me'. It was a German high explosive bomb, dropped from 14,000ft. Fortunately for Joe it failed to explode.

Joe took a new job at the start of 1941, working as a messenger for a shop in the City which made military hats. Each day Joe took the bus across the Thames and into work, a dreary commute that left him little time to see his pals. Saturday was a later start than weekdays, so he was making his way up the Walworth Road for his weekly wash at the Manor Place Baths. 'You had these cubicles with their own private tubs,' he remembers, 'and the attendant would fill it to the top and then if you wanted it hotter or colder, you'd shout out "more hot water in Number 10!" and he'd come and add a bit more hot water. It only cost a few pence and it was our chance to have a bath in peace and quiet.'

If Joe had turned left out of the baths once his ablutions were over, walked along Penton Place for a couple of hundred yards, he would have been on Newington Butts with the Elephant & Castle a five-minute walk north and Kennington Park Road to the south. Right in front of him would have been the church of St Mary's Newington, built of Kentish ragstone. The interior roofs of the nave and chancel were of hammer-beam construction, the height of the nave from floor to roof ridge being 70ft.

It was an impressive church even without the steeple that the architect neglected to add, but perhaps he just had uncanny foresight because one of the more extreme precautionary measures of a twitchy British Government in 1940 had been the paring down of exceptionally vertiginous church steeples for fear they might topple down and crush people during bombing raids.

The present vicar of St Mary's was the 29-year-old Christopher Veazey. His father had been a parish priest in Camberwell since 1894 and he had inherited his father's rectitude and earnestness. Christopher had needed someone to loosen his chains of righteousness and he found her in a Wimbledon boarding house, where, on 11 September 1937 he became a lodger. Joan, the 24-year-old daughter of the landlady wrote in her diary that evening: 'The curate arrived today pushing a large barrow full of his books and worldly goods.' In those first few days the young vicar was terrified of Joan. Unsettlingly good-looking, with silky black hair and eyes that made him stammer and blush, it was her personality that swamped him. Joan had inherited from her Australian mother a bubbling irreverence and humorous disregard for stuffy English conventions. When she told Christopher over supper one night that she worked as an artist's model in her spare time he practically choked with horror. Joan liked to goad him at the dinner table by revealing she 'hadn't been inside a church for years'. They inhabited different worlds, he with his strict Victorian upbringing, and she with a family who saw nothing wrong in displaying affection. But within a month Joan was confiding to her diary: 'He is good looking with lovely dark hair, which has a mischievous wave in front, which he tries hard to cover. He has brown eyes which seem to look right through me.'

Joan began to fall in love with Christopher, this uptight man whose kindliness overwhelmed her. At her insistence he took her with him on one of his regular evening visits handing out food to London's down and outs. She was left numb by the misery of these people's existence. 'What is wrong with this country of ours,' she asked her diary, 'that we should allow young men to sink into such depths of despair?'

By the end of 1937 the pair were engaged to be married, and by the end of 1938 he was her equal in teasing. 'Darling,' he would say to Joan, 'I don't know which I prefer: kissing you or smoking my pipe.' They married in Camberwell in September 1940, and the wedding reception had been more memorable than most. As the happy couple posed for photographs, the Battle of Britain was approaching its denouement in the skies above. 'Without any warning,' Joan wrote in her diary, 'a German plane flew in low over the house and released a bomb. I shouted to the photographer who was under the black cloth trying to take the photo, "Look out, he's dropping bombs!". The man shouted back from the cloth, "Never mind about the bomb, just smile!".'

For the first few months of their married life they had lived at the rectory with Forest FitzGeorge and his wife, but since January they had been ensconced in their own flat further down Kennington Park Road. 'It wasn't really a flat,' remembers Joan. 'It was a two room place with a toilet downstairs which we shared with another family.' But it was their own flat and they were happy. Not long after, Joan had discovered she was pregnant and they were even happier.

At 10 a.m. on Saturday morning 10 May, Christopher was sitting at his desk preparing his Sunday sermon. Joan was tempted to go shopping at the Elephant & Castle, and perhaps sneak into the Trocadero for the Saturday matinee, but she was worried that some of the parishioners, those ones she found to be a 'bit sniffy', might not approve. 'I wanted to be like a church wife,' Joan says, 'but I wasn't like one. I liked dancing and stage shows, and I wasn't very religious.'

IT was a perilous route, from St Mary's Church to the Elephant & Castle, even though it was only a few hundred yards. It was noisy, too, with trams going up and down Newington Butts on their way to and from Tooting and Streatham. The din reached a crescendo at the Elephant & Castle with trams, buses and cars vying for space underneath the overhead railway line that ran north to south parallel to the Walworth Road. A pedestrian subway had been built under the road junction in 1911 to help people get across the Elephant & Castle without being hit by a bus, knocked down by a tram or trampled by a horse and cart. The children were delighted. 'All the walkways led down to the middle of the subway,' remembers Emily Macfarlane, 'and there was a man there whose job it was to point people in the right direction and generally keep the place clean and tidy. Me and my mates used to torment him. We'd each go to a different entrance and then take it in turns to run down, shouting and hollering, and it made such a noise with the echoes. "I'll get you," he'd cry and as soon as he started chasing one lot the other lot would run down shouting and screaming. Poor man!'

ANYONE heading north on a tram at 10.30 a.m. on Saturday 10 May would have noticed more activity than normal at Waterloo railway station. People coming, people going, the end of some love affairs, the start of others. The concourse at Waterloo teemed, but 14-year-old John Fowler had found his mum and dad. Tears and hugs all round, so different from eighteen months ago when he had been evacuated to Haslemere in Surrey. Separated from his parents, John had lost his sister, Joan, because of an administrative bungle. 'We should have both gone to Haslemere but they put Joan on the wrong train and she ended up in Sturminster Newton in Dorset.'

In the year and a half since his evacuation John had grown into a young man. Life as a farm labourer had sheared off his city flab and his eyes flashed with a healthy zest. In contrast the faces of John's parents, despite being wreathed in smiles, showed what they had lived through in the last eight months. 'They never said anything about the bombing in the letters they wrote to me,' remembers John, 'but I just had to look at them to know how bad it had been.' John was back in London for his sister's thirteenth birthday party. Today was the chance to catch up with his friends who were still in Peckham, as tomorrow the family was heading to Sturminster Newton for a snatched reunion. Foremost in John's mind, as he waited with his parents for a tram outside the station, was Rose, his 15-year-old cousin. She had refused to be evacuated from her house just down the street from John, preferring to stay and work for a City firm. The two were good friends and John was hoping he might be able to take her to the cinema in the evening.

On the other side of Waterloo Road from the tram stop was the fire station, 101 Waterloo, where 27-year-old Auxiliary Fireman Fred Cockett had reported for his 48-hour shift an hour earlier, bang on 9 o'clock. It was one of Fred's characteristics that he was never late, despite the 7-mile cycle from Brockley. Diligence and good humour were Fred's other two outstanding qualities, and his humour manifested itself best in his story-telling. He could spin a yarn, Fred, much to the amusement of the two veteran station officers, 'Pinky' Petit and 'Tiddly' May. They had been brought out of retirement to command 101 Waterloo* and they had appointed Fred as their driver. 'They had a staff car,' recalls Fred, 'but they never wanted to drive so I was chosen. Pinky and Tiddly were both widowers so they moved into the station, gave me their ration books and a £2 float and told me to look after the provisions.'

Fred's funniest anecdotes centred on his pre-war career in the

*'Waterloo Fire Station,' says Fred, 'was a "Ghost Station", outside the control of any particular local authority. We were a back-up so that if any other station got bombed we could replace them.'

catering industry, a strange choice of profession in itself for a man whose great uncle had been the skipper of a Lowestoft windjammer and whose father had driven tanks in the First World War. Fred had attended Hotel School in London, then spent six months working in a Paris restaurant in the mid 1930s. He returned to work in the members' dining room of the House of Commons, serving dishes and clearing plates as Britain's leaders led the country inexorably towards war with Germany. 'I remember serving at the cabinet table not long after war had been declared,' says Cockett. 'Sir John Anderson, who was Home Secretary, was talking to someone else about setting an intelligence test for some service or other. Anderson said "Let's give them a series of numbers and they have to work out the next number in the sequence".' For the next few minutes the two politicians devised a puzzle as Cockett waited on them discreetly. 'They came up with something and John Anderson saw me working it out under my breath. "What do you make it?" he asked. I gave him the right answer and he turned to his companion and said "There you are, you've got your first recruit!". At this point I had to tell him I had just volunteered for the fire service.'

Cockett had enlisted in the Auxiliary Fire Service shortly before war was declared, rallying to the cause out of patriotic duty rather than any compulsion to be a firefighter. 'There was an exhibition of auxiliary emergency services in County Hall,' he recalls, 'and I had had most afternoons off so I went along and asked the services if they did any afternoon training. None of them did except the Fire Brigade so I volunteered for them.'

Like all part-time Auxiliary Firemen and women, Cockett was mobilized on Friday 1 September. He reported to Lambeth HQ with his knife, fork and spoon, two blankets and shaving utensils, leaving behind his wife and baby. Uniforms were distributed, 'dungarees, rubber boots and a cap because we didn't have proper uniforms then'.

The double-breasted tunic, leggings and steel helmets came later, by which time the Auxiliary Firemen (AFS) had been trained into the ground by the regular firemen. They could tie every conceivable knot with their eyes closed, they were intimate with their stations' ladders, could come down a ladder with a colleague on their back and they had perfected escape techniques from the third floor of the drill tower. The culmination of the training was a jump from the drill tower. 'You had to jump from about 30ft into this sheet,' says Cockett. 'It might not sound very high but when you were standing up there the sheet looked like a penny.' It was the men holding the sheet who were in more danger than the jumper. 'The rescuers had to tuck their chins in and on no account look up. If you did, the jerk as the jumper hit in the sheet could whip back your head and break your neck.' All this for £3 a week, and who-do-you-think-you-are stares from the regular pre-war firemen, many of whom resented the newcomers.

Away from the training ground, the AFS were put to work on more mundane tasks: sandbagging the schools that had been converted into sub-stations for the duration of the war; painting all vehicles a dull grey colour and keeping the station spotless. As the months slipped by, and still the German bombers didn't come, the public began to pour scorn on the men who stayed at home while the three armed services were 'doing their bit'. 'We came in for a fair bit of abuse,' remembers Cockett. 'They called us army dodgers and the like. We didn't like it because we knew how bloody hard we were working.'

As the remains of the British Expeditionary Force scrambled away from Dunkirk, public hostility grew, whipped up by some politicians. In a debate in the House of Commons one Member of Parliament accused the Civil Defence of being infested with 'parasites and slackers'. Soon firemen in uniform were being refused entry into London restaurants; others were pushed and shoved in the street. Graffiti was daubed on the walls of many auxiliary sub-stations: 'Cut

Price Firemen, £3. Not worth it'. By August 1940 the slow trickle of some of the 20,000 Auxiliary Firemen into the armed services had become a torrent.

But the Luftwaffe flew in to the firemen's rescue. Within a week of the start of the Blitz in September the firemen became more popular than Winston Churchill himself. The newspapers championed them as the 'Heroes with Grimy Faces'; restaurants that had shut the door in their faces now grabbed them off the streets and fed them free of charge. The irony of it all amused the Air Raid Wardens, who corrupted one of Rudyard Kipling's famous poems in tribute to the fickleness of it all. 'Oh it's "Wardens this!" and "Wardens that!" and "Wardens, you're a flop"/But it's "Thank you, Mr Warden" when the bombs begin to drop.'

AFTER arriving at Waterloo Fire Station, one of Fred's first tasks, even before the half hour game of volleyball that constituted the daily physical training, was to nip across the road and stock up with provisions from the shops and market in The Cut.

The Old Vic Theatre stood on the corner, an incongruous sentinel at the entrance to one of south London's most bustling streets. The Cut joined Waterloo Road to Blackfriars Road, and along it were shops and market stalls selling all imaginable produce: Boyd the grocer, Crisp the baker, Cox the butcher, Phillips the tobacconist, Bronstein's the tailor and Wood's Eel and Pie shop, where the live eels were displayed in tanks at the front of the shop. Customers picked the one they fancied and the owner fished it out. 'When they stewed the eels,' says Joe Richardson, 'they didn't throw the water away, they put parsley in it and a bit of flour to thicken it up and that was the sauce. With a bit of mashed potato it was beautiful.'

Waterloo Railway Station lay on the inside bend of the Thames as it angled sharply south towards Vauxhall. A couple of hundred yards north of the station was the bridge that bore its name, though in 1941 a new bridge was under construction alongside the old one. To the south of the station was Westminster Bridge, linked to Waterloo Bridge by York Road, a shabby street of assorted warehouses, store-rooms and offices that ended at the southern side of Westminster Bridge.

From here one could peer south and see, on their right, the nine separate pavilions of St Thomas's Hospital strung out like a necklace along the Thames, and on the left, through the trees, the rough Kentish rag of the four-storey Lollards' Tower, the highest point of Lambeth Palace, residence of the Archbishop of Canterbury, which once had been accessible from the river.

St Thomas's, like every London hospital, had steeled itself well for the Blitz. In 1939 it was designated a Casualty Clearing Station with 330 beds, of which 200 were held in reserve. Sixteen of the upper wards were closed and their beds distributed among the other wards. The main operating theatres on the top floors were moved to the basement, where previously great quantities of linen had been stored. All out-patient services were suspended, though the tuberculosis and venereal disease sections remained open. The training school was evacuated and student nurses were transferred to other London hospitals. Those staff that remained helped Civil Defence workers build a protective wall of 30,000 sandbags around the pavilions, or blocks as the nurses called them, of St Thomas's.

Behind Block Five was the hospital's newest building, Riddell House, which had opened in 1937 and now was home to nearly a hundred members of the Ambulance Train Service and the River Emergency Service. There was frenetic activity inside Riddell House on the morning of 10 May 1941; people cleaned and scrubbed and swept and cast anxious glances at their watches. The Archbishop of Canterbury was due to arrive in a few hours to lead the hospital in prayer, the first time Riddell House would be favoured with his presence. The place had to be worthy of His Grace.

For anyone blessed with a vigorous constitution the walk from Lambeth Palace to Riddell House was no more than a five-minute stroll. But Cosmo Lang was a frail old man whose spirit rose and fell like the flickering of a candle. He had suffered paroxysms of misery in his ecclesiastical life; first in 1914 when as Archbishop of York he had spoken out in those first frenzied weeks of war at what he regarded as the demonization of the German people. The newspapers savaged him, accusing him of disloyalty and triggering an attack of alopecia that aged the Archbishop prematurely. He had finished the war haggard and bald, save for a few wisps of grey hair over his ears.

In 1937, nine years after his appointment as Archbishop of Canterbury, Lang had presided over the abdication of Edward VIII so that Edward could marry Wallis Simpson. When he crowned George VI king in Westminster Abbey on 12 May 1937 it was, Lang said, 'like waking after a nightmare to find the sun'. But the constitutional crisis had taken its toll. His chaplain, and one of his few close friends, Alan Don, begged Lang to rest. 'Your physical frame is showing signs of wearing out,' he told him.

The Archbishop had heeded Don's advice, and slowed down, but the spectre of Germany returned to haunt him in 1940. Bombs landed in the Palace gardens during the first week of the Blitz uprooting some of the fig trees that were descendants of those planted by Cardinal Pole in the 1550s, the last Catholic Archbishop. On 17 September Lang left for the Old Palace at Canterbury, leaving Alan Don in charge. He took his leave reluctantly, aware that Lambeth Palace had been the London seat of Archbishops of Canterbury since the early thirteenth century. With him went the most valuable of the library's 50,000 books and manuscripts, including the first three editions of Thomas More's *Utopia*. A few days after Lang's departure, a bomb screamed through the big oriel window and exploded in the drawing room, wrecking three bedrooms.

Lang returned sporadically to the palace over the next eight months, but Alan Don was a reliable man to have overseeing matters. He made daily reports to Lang, telling him about the bomb that had exploded in Church House in October; that two hundred local people still used the crypt, the oldest part of the palace, as an air raid shelter, and of his own appointment as rector of St Margaret's Church, Westminster, in January 1941.

On 28 January Don – by now a canon – and his wife, Muriel, moved into 20 Dean's Yard, an attractive courtyard at the south-west corner of Westminster Abbey. No. 20 had been built in the late fourteenth century, but was now an amalgam of different styles: the brickwork of the upper floors and the Gothic windows dated from the eighteenth century, the grisaille decoration in the first-floor rooms was two hundred years earlier and the roof of another room was an original survivor from the 1380s.

From his new abode Don could see the contusions on the 900-year-old Abbey; many windows had been blown out but apart from a few autographs from bomb shrapnel on the walls there was nothing irreparable. Inside 60,000 sandbags protected the tombs of England's greatest: Edward the Confessor behind the High Altar, Henry V under his Chantry Chapel and close to the west door, the grave of the Unknown Warrior. Movable treasures such as paintings and manuscripts had been evacuated to the country. The Coronation Chair on which George VI had been crowned was in Gloucester Cathedral. The memorials and artefacts that remained were guarded by the Abbey's own Civil Defence team. During the dragging months of the Phoney War, there had been lectures – from first aid to firefighting – in the Norman Undercroft and a warden's post was installed in the Pyx Chapel in the East Cloister.

A stone's throw south of the Abbey, against the east wall of the

College Garden, there was an oblong air raid shelter with a concrete roof over 8ft thick and capable, so the Dean of Westminster, Dr Paul de Labillière, was assured, to withstand a direct hit from a 250kg high explosive bomb. A hundred yards north of the Abbey was Alan Don's church, the fifteenth-century St Margaret's, which had been adopted as the parish church of the House of Commons in 1614. Samuel Pepys married 15-year-old Elizabeth St Michael in St Margaret's in 1655 and was a regular worshipper thereafter.

When Canon Don took up his new appointment in January 1941 he stood on the level lawn outside the west entrance and examined his church. The stone tower and the low-roofed nave were untouched, Don noted in his diary, but a high explosive bomb during the early Blitz had left 'most of the windows and vestries irretrievably shattered'. Another bomb in September 1940 had sliced open the pipe that carried the wind to the organ from the blower in the tower, the same organ that had accompanied Winston and Clementine Churchill down the aisle on the day of their wedding in 1908 with the groom wearing what was described in a fashion magazine of the time as 'one of the greatest failures as a wedding garment we have ever seen'.

HAVING spent a week with Archbishop Lang in Canterbury Canon Don had returned to find yet more damage to St Margaret's following the heavy air raid of 19 April. And there were other pressing issues requiring his attention. Church services had resumed at Whitsuntide after a winter shutdown caused by the bomb damage. In the interim the People's Warden of St Margaret's, John Rathbone, Member of Parliament for Bodmin Division and RAF Flying Officer, had been killed in action. Who to appoint as his replacement? In February Don had persuaded Leonard Eaton-Smith, the Mayor of Westminster, to stand. 'I must get him elected by hook or crook,' he wrote in his diary. 'He would be a real acquisition.'

Don had been impressed by Eaton-Smith's integrity and industriousness in the seven months he had been in office. Though he was the son of a Liverpool merchant, the 67-year-old Eaton-Smith had lived in Westminster for forty years and his accession to the Mayoral office was the natural culmination for such a respected figure. 'My year of office may be strenuous, difficult and exacting,' he said in his acceptance speech, 'but I say let us maintain our cheerfulness.'

Canon Don looked at his watch on the morning of Saturday 10 May. It was approaching 11 a.m. In a couple of hours he was having lunch with Mr Wilmshurst, St Margaret's senior sidesman, to discuss the possibility of Eaton-Smith as the new People's Warden. He was confident Mr Wilmshurst would agree to his nomination.

CANON Don was something of a hit with the firewomen of 'A' Division Headquarters in Little Dean's Yard, a cobbled court at the east of Dean's Yard that was part of Westminster School. 'He looked like one of those ghosts you see in pictures,' recalls Reenie Carter, a 20-year-old firewoman at the time. 'He had a long drawn face, quite white, but he was very jolly. Next to us was the Abbey Garden, which was closed to the public, but Canon Don gave us the key to the gate so we could sit in there when we were off duty. Sometimes if it was warm enough we would take our blankets and sleep in the Garden.'

Don had cause to be grateful to Westminster Fire Service, blessed as he was with two sub-stations within shouting distance. 'One station was in the choir school,' says Reenie, 'and the other was in Westminster School in Great College Street.' Reenie and the girls were stationed in the control room of HQ, 'a brick surface air raid shelter in Little Dean's Yard', sandwiched between the Abbey Garden and Dean's Yard.

Reenie's comrades were a mixed bunch, 'from society women to the roughest of the rough … one firewomen had a maid at home and took breakfast in bed each morning but when she came to work she took turns scrubbing the stairs with everyone else'. Reenie's closest friends were Josephine Kam-Radcliffe, Margot Seymour-Price and Hettie Goodchild, and they all had the run of Westminster School, which had decamped en masse to Herefordshire. With them went the choir school, their removal being 'a disaster of the first order', in the opinion of Dr Jocelyn Perkins, the Abbey's 71-year-old Sacrist. It was the first time since the Restoration of 1660 that the Abbey had no choristers. 'The collapse of congregations and collections is simply pitiful,' said Perkins. 'To walk across the Little Dean's Yard with its empty buildings, devoid of all traces of the boys, gives one a fit of the blues ... no individual or group is to blame, except Hitler.'

The fire service made hay in the deserted school. 'We used it as the living quarters,' says Reenie. 'The ground floor was used as "A" Division's Administration Offices and the headmaster's drawing room was turned into a dormitory.' The table-tennis tables and snooker tables left behind were avidly used by firemen and women, and sometimes Canon Don popped his head round the door to issue a sporting challenge.

There were other perks to the job, too. The previous Christmas Reenie and Hettie Goodchild had stood in Little Dean's Yard, 'on a brilliant moonlit night', listening to an improvised choir singing carols. As the voices drifted across the Yard, Reenie turned to Hettie: 'They sound like angels, don't they?'

3

PEACE had come momentarily to Westminster, and few places in London needed a respite as much as the Houses of Parliament. The parliamentary foundation stone had been laid in 1840, but Parliament's gentle contentment had ended one hundred years later with the Blitz. In eight months the Houses of Parliament had been hit eleven times. Miraculously, the only severe damage had been caused by the high explosive bomb that fell in the Old Palace Yard on 26 September. Otherwise, the buildings looked as they did in 1852 when Queen Victoria and her husband, Prince Albert, had opened Parliament. The Gothic splendour, the creation of Charles Barry, still took one's breath away as one walked over Westminster Bridge,

Running alongside the river was the Members' terrace, 678ft long and recently adorned with a machine-gun nest, in case the Germans chose to invade up the Thames. It was hoped that the barricades of barbed wire in Parliament Square might deter German paratroopers from dropping in to capture Churchill. The last line of defence was a W. H. Smith's bookstall in the square (in reality a pillbox with a machine gun manned by the Westminster Home Guard).

At the southern extremity of the Palace was the 323ft Victoria Tower, inside which the parliamentary records were kept. The Clock Tower (6½ft smaller than the Victoria Tower) was at the north end.

Its 13-ton bell had first chimed in 1858 and was named Big Ben after Benjamin Hall, the Commissioner of Works at the Whitechapel Bell Foundry where it was cast. Since 1940 its chimes were heard up to forty times a day on the BBC Overseas Programmes and to the Americans it was 'the Signature Tune of the British Empire'.

The Houses of Parliament rested on a vast bed of concrete, over 10½ft thick, an insurance against the unstable ground on which it had been built. Here in Parliament on 13 May 1940, Churchill had risen in the Chamber of the House of Commons to make his inaugural speech as Prime Minister. 'I have nothing to offer,' he told the House, 'but blood, toil, tears and sweat.'

Since the Blitz had begun Churchill's opportunities to hold sway in the Commons Chamber had been restricted. It had been thought advisable in November 1940 to move the Commons temporarily to the more sturdy Church House, headquarters of the Church of England, just 500yds away facing Westminster Abbey. The Commons had returned to the Chamber in January – 'we missed our warm panelling and spacious lobbies, our Library and Smoking Room,' said A. P. Herbert – but two heavy raids in spring had sent the politicians scuttling back to Church House. They were back home by 7 May, however, almost one year to the day since Churchill's maiden speech as Prime Minister. Now he was facing a vote of confidence in his premiership. He survived with an overwhelming majority, 'Ayes 447, Noes 3', but the future for Britain and Churchill looked grim. Germany now occupied Yugoslavia and Greece, Rommel was menacing British forces in the Middle East and Luftwaffe bombers had spent the past week laying waste to Liverpool (where sixty-nine out of 144 berths in the vital docks were destroyed), Glasgow and Humberside. Churchill had ended the debate on his leadership in his own inimitably bullish way. 'When I look back on the perils which have been overcome, upon the great mountain waves in which the gallant ship has driven, when I remember all that had gone wrong, and remember also all that has gone right, I feel sure we have no need to fear the tempest. Let it roar, and let it rage. We shall come through.'

The afternoon of Wednesday 7 May had been one of robust debates all round. Dr Haden Guest, Member for North Islington, had clashed with the unpopular Herbert Morrison, the Minister for Home Security, over the issue of firewatching. Was he aware, Guest demanded, 'that it has become a custom for people with means of transport and wealth to leave the area each night, thereby leaving their houses and business premises unprotected and placing an extra burden upon those who remain who take up the duty of firewatching?' Morrison replied that he was aware. Then was he inclined to make it 'compulsory that these people should either remain or make provision for firewatching of their homes or business premises?' Morrison prevaricated. 'There are many difficulties,' he said, 'in compelling people to come back from a distance.' Guest asked his final question of the debate: 'So is there any obligation on owners of property at all with regard to firewatching?' 'No' said Morrison. 'That,' retorted Guest, 'is a scandal.'

Nowhere was it more of a scandal than in Westminster and Chelsea. The moment war had been declared the occupants of large swathes of these prosperous boroughs had fled to the country. By March 1940 the population of Chelsea had dropped from 57,000 to 36,000 and a third of rateable properties stood empty. In Belgrave Square, thirteen of the forty-five houses were placed on the market.

At first the working class had treated the exodus with humorous derision; but their mood turned ugly when the bombs started to drop. The problem, explained the *Westminster & Pimlico News,* was that the water rates in these empty houses 'are unpaid and so water is cut off.

When a fire does occur it's useless for the neighbours to try and get the fire under control because there's a lack of water.' In the same article one of Mayor Eaton-Smith's councillors pleaded for volunteer firewatchers, but by May 1941 only 40 per cent of Westminster residents had registered under the compulsory Firewatching Order of January 1941.*

The migration to and from Chelsea altered the area's character. The rich were replaced by the more stout-hearted, in particular military men and members of the Civil Defence. One young soldier posted to the Duke of York's barracks in Chelsea was Harry Beckingham from Manchester. 'There was an YMCA just off Sloane Square,' he recalls. 'We used to go there to play table tennis and when we were paid we walked up to Chelsea and went to one of the clubs. I remember seeing Phyllis Dixie once or twice, a famous stripper who used to come on stage behind a couple of small fans. I think she got banned, but all the boys liked her.'

Beckingham was 20 when he arrived in London in the first week of the Blitz, one of fifteen men in No. 35 Bomb Disposal Section, Royal Engineers, under the command of Second Lieutenant Godsmark. Beckingham was tall and slim, with an intelligent face tanned by a summer of relentless training. He had the confidence of youth but, crucially, he wasn't over confident. 'I'd joined the territorial army in June 1939 and was called up on 1 September. My Field Company was sent to France but because I'd just turned 19 I was left behind in Manchester.' Sapper Beckingham was still kicking his heels six months later. 'Then one day the sergeant-major asked me if I'd like to go on a course.' Desperate to break out of the tedium of camp, Beckingham

* After the City was firebombed on 29 December 1940 Herbert Morrison introduced the Fire Precaution and Business Premises Order in January. In theory it was compulsory for men between the ages of 16 and 60 to do forty-eight hours firewatching a month, i.e. to occupy empty buildings at night and guard against incendiary bombs. In practice, however, there were so many loopholes for the working man that 75 per cent of those eligible were exempted.

went down to Sheffield to start a course he knew nothing about. 'I discovered it was a bomb disposal course.'

Throughout the summer of 1940 Beckingham became one of a small band of Royal Engineers trained in the art of defusing bombs. The teaching was rudimentary and it was often inaccurate. 'We were told that when bombs dropped they would be lying on the surface, so we practised putting a sandbag wall round the unexploded bomb with a small crawl hole. On top of the wall we lay a sheet of corrugated iron with four more layers of sandbags on top of that. Once that was done we crawled through the hole, put a 1lb charge of gun cotton on the bomb and detonated it.'

Once in London Harry and his section (now incorporated into No. 35 Company with a strength of thirty-two men) soon realized that unexploded bombs 'didn't lie on the surface but rather 20ft under'. And occasionally in the most inconvenient of places. 'A stick of unexploded bombs landed in a cemetery in Leytonstone, so we had to start digging them up. The bodies were stinking to high heaven and the only way we could kill the stench was to pour creosote round the holes. We removed the bodies with shovels, but they disintegrated as soon as they hit the air. We put them to one side, got down into the ground to defuse the bombs, and then shovelled them back in afterwards.'

From 10 to 30 September 1940, No. 35 Bomb Disposal company defused 470 unexploded bombs (UXBs). Incredibly, through a confection of skill and good fortune, they suffered no casualties. But they lived dangerously. 'One time we had spent a couple of hours working on this 250kg bomb in a back garden in Ilford,' remembers Harry. 'We'd attached a rope from the bomb to our lorry to try and shift it but it wasn't coming, so we decided to pack up for the night and come back in the morning.' As they drove away the bomb detonated, taking most of the back garden with it.

For nearly two months Harry's company defused bombs. They worked from 8 o'clock in the morning to 8 o'clock at night, stopping only when it was too dark to see what they were defusing. 'We got up in the morning, had breakfast, climbed into the three tonner and went off to a job. I can't ever remember being worried, but I was young.' Other men, those who lacked Harry's phlegm, or were too old to believe any longer in the indestructibility of youth, did suffer the psychological pressures of the work. 'There was a sergeant who got stressed out,' says Harry. 'And it wasn't long before a rule was introduced allowing men to transfer out of Bomb Disposal once they'd done six months.'

Once the initial fury of the Blitz had abated Harry's company was posted to Stanmore in Middlesex. The bombs were no longer dropping every night but there was a backlog of dozens, scores, hundreds of UXBs to defuse. In March 1941 No. 35 Bomb Disposal company moved to Chelsea and the two heavy April raids gave them plenty of work in the West End. 'The trouble with London,' says Harry, 'is that it's all basically built on blue clay, so bombs slid down to depths of 20ft or more.'

THE West End Harry encountered in the spring of 1941 was jaded and tatty. It no longer resembled the glittering, garish district Harry had seen in pictures before the war. The neon signs and electric messages of Piccadilly Circus were gone, and Eros – whose statue had bathed in their illuminations – was clutching his bow behind a giant wall of sandbags. Other West End statues had been removed for safekeeping. Also absent were the fountains of Trafalgar Square, and in their place were red brick air raid shelters, windowless and intimidating. Nelson remained on his column but the statue of King Charles I astride his horse had been penned inside a wall of brick and timber. People well acquainted with Trafalgar Square before the war noticed something else too; there were hardly any pigeons left.

What colour there was came from the Government posters plastered on every available space – 'Dig For Victory', 'Beat Firebomb Fritz – Join the Fire Guard' and 'Be Like Dad, Keep Mum – careless talk costs lives'* were just a few – and from shops whose windows had been blasted. In place of glass the owners had erected plyboard facings. Small holes had been cut in the middle so customers were able to see the window displays. All around the plyboard were painted vivid and exuberant depictions of the shop's wares.

By 10 May 1941 such shops were commonplace in the West End. The district had been knocked about a month earlier when, on the night of 16 April, two hundred high explosive bombs landed in Westminster. A shelter in Leicester Square took a direct hit with many casualties, while parts of the Strand were pulped. The Shaftesbury Theatre was destroyed and the 200-year-old Stone's Chophouse in Panton Street, renowned for its steaks and grilled herrings in mustard sauce, was reduced to a sad ruin.

Panton Street was at the centre of Constable Ballard Berkeley's beat. He knew the area well and he trod the pavements as confidently as he had the theatre boards before the war. 'I knew the West End backwards, all the theatres, nightclubs and restaurants.' Ballard had made his London stage debut in 1928, aged 23, in *The Devil's Host*. A couple of years later he appeared in his first film, *The Chinese Bungalow*, in a cast that included 'matinee idol' Matheson Lang. The 1930s had provided Ballard with a steady if unspectacular flow of work on stage and screen, but he was still waiting for the part that would make his name. He hadn't found it in the 1939 RKO Radio film *A Saint in London*, the last movie Ballard made before the outbreak of war.

* This last slogan provoked a sharp exchange in the Commons on 6 May 1941 between Edith Summerskill, Socialist Member for Fulham, and Duff Cooper, Minister for Information. Summerskill said that the phrase 'Keep Mum' was 'offensive to women', to which Cooper replied: 'I cannot believe the irritation about this poster is very profound or widely spread.'

Too old for the armed forces, Ballard volunteered for the Metropolitan Police as a special constable and was posted to West End Central Station. Ballard, loose-limbed and light-hearted, had the voice of an actor, deep and sure, like the draw of a bow across a double bass. The prostitutes and petty criminals of the West End were proud to have their collars felt by the famous policeman.

Ballard's beat was small and shaped like a rectangle, with the Haymarket, Whitcomb Street, Pall Mall East and Coventry Street as the four sides. Coventry Street linked Piccadilly to Leicester Square. It was in Coventry Street on 8 March 1941 that Ballard was confronted with real death, so different from the movie version. That night, two 50kg bombs exploded in the Café de Paris, a restaurant frequented by well-heeled types. The first bomb burst above the stage where 'Snakehips' Johnson was leading his West Indian band in *Oh Johnny*. All the band except Leslie 'Jiver' Hutchinson were killed. The second bomb landed on the dance floor. 'In such a confined space the blast was tremendous,' recalled Ballard. 'It blew legs and heads off and exploded their lungs.' The emergency services were quick to arrive, as were the looters. 'One hears a lot about the bravery during the war,' said Ballard, 'but there were some also very nasty people … these people slipped in pretty quickly and it was full of people – firemen, wardens, police – so it was very easy to cut off a finger here [to get a ring] or steal a necklace, and it did happen.'

By 10 May the carnage at the Café de Paris had been forgotten. London's dance halls were packed with a clientele swinging to Glenn Miller's 'Tuxedo Junction', the latest dance craze doing the rounds. Inside the Lyons Corner House on Coventry Street people sipped their tea and read the morning's papers, each one full of advertisements promoting that evening's live bands and West End shows. Most theatres had followed the instructions of the Government and shut up shop the moment war was declared. But when the Luftwaffe hadn't appeared by December the theatres began to reopen their doors. They remained open even when the Blitz did start, although performances were brought forward to allow the audience to get home before the bombers closed in.

The most popular show in town in May 1941 was George Black's *Applesauce!* at the London Palladium off Oxford Street. It was Variety at its best, bringing together the most popular acts of the time: comedian Max Miller and his double entendres, Florence Desmond and her witheringly accurate impersonations, among them Marlene Dietrich, Greta Garbo and Tallulah Bankhead, and the new 23-year-old singing sensation, Vera Lynn. 'The show had started off in the Holborn Empire,' recalls Vera, 'but that took a direct hit in 1940 and we moved to the Palladium.'

Vera had been a band singer since 1935 but it wasn't until she joined the Ambrose Band in 1937 that she hit the big time. 'They were the number one band in the country and I was happy. I'd achieved what I'd wanted to do, which was to sing with the best. So when war came along I thought "Well, bang goes my singing. I'll be working in a factory from now on". I didn't think for one moment of building a career.'

But when the authorities realized the people wanted entertaining they looked around for a female star, and Vera Lynn fitted the bill. She was young and pretty with a voice that had a unique, indefinable, quality to it. When Vera sang it was as if she was singing just for you. No one else, just you. 'I didn't work on that,' says Vera, 'it was just the way I felt each song and what it was for.'

Vera's star was on the up but her feet stayed on the ground. There were no tantrums or tears or tiffs with fellow performers. She lived at home with her parents in Barking and she had no chauffeur to drive her to and from the Palladium. Vera drove herself, with a tin helmet perched delicately on her head, in a green canvas-roofed Austin 10 she had bought for £200. 'Once I'd been going home from a show when the air raid siren went as I drove through Poplar. "God, what do I do?" I thought, but the only thing to do was to go on, and that's what I did.'

At 11.30 a.m. on 10 May 1941 Vera was about to leave home for the drive in to the Palladium. There were two shows that day, the matinee beginning at 2.30, and the evening show at 5.15. The journey for Vera was straightforward, especially as there had been no raid for three weeks. The diversions and detours of past months were gone and, though the craters weren't, they'd been filled in and all roads were passable. The commute took her through some of London's worst hit areas: East Ham, West Ham, Poplar, Aldgate, Holborn and along Oxford Street to the London Palladium. So far, only the Palladium had evaded the bombs.

Holborn in particular was a favourite quarry of the Germans. Bombs had shattered the borough with frightening regularity since 7 September. On 26 September Britain's answer to Lord Haw-Haw, Sefton Delmer, had been entertaining in his flat in Lincoln's Inn. Delmer, born in Berlin to a British woman, had fled Germany at the start of the war and in the summer of 1940 began broadcasting to the Germans on the BBC. He and his wife were terribly well connected, and Delmer liked to make the most of his connections. Among those present at this dinner party were Martha Huysmans, the daughter of the Belgian Prime Minister, Prince Bernhard of the Netherlands and Ian Fleming, Personal Assistant of the Director of Naval Intelligence at the Admiralty and later to win fame as the creator of British agent James Bond. 'I had just persuaded Bernhard as he was putting on his coat to leave,' Delmer said later, 'that he must stay for a last brandy, when what seemed an end-of-the-world explosion sent us all sprawling. The building heaved as in an earthquake ... I walked over to the entrance hall and opened the door. "It's nothing," I said grandly ...

We lit some candles and continued to gossip and drink champagne as if nothing had happened.'

Delmer's heroic indifference to the bomb had gone down well with his *Daily Express* readers (for whom he worked as a journalist), but the first fireman on the scene remembers it differently. 'We rushed over there,' says Lew White, 'and the front of the building had been smashed completely. I helped out Prince Bernhard who was looking for his girlfriend. He was very grateful and the next morning came to the station and presented us all with china mugs.'

Lew White was an auxiliary fireman at station A1Y in Stone Buildings, Lincoln's Inn. The rectangular courtyard of Stone Buildings had history. Built in the 1770s, they were where William Pitt the Younger had his Chambers and the Inns of Court Volunteer Rifle Corps – first formed to defend Queen Elizabeth – had been based at No. 10 for over fifty years. Lew's sub-station was in No. 7, at the southern end of the courtyard and opposite No. 10.

The 27-year-old White came from a long line of north London Jewish butchers, although Lew preferred to spend his Saturdays watching Arsenal rather than going to the Synagogue. Since leaving school at 14, Lew had worked in Smithfield Market, handling meat carcasses that were bigger than his 5ft frame. He remembers it as a hard life. 'Smithfield opened at 4.30 each morning and in the winter it was very cold. Fortunately the pubs opened early so we went in there and drank a couple of rum and milks to keep us warm.' In his early twenties Lew spent his days off either following the Arsenal or in Hampstead with his best pal, Michael. 'We were looking for girls, not a wife, but one Sunday I met this beautiful girl called Frieda. We married in 1937 and our first son was born in June 1939.'

A month later Lew joined the Auxiliary Fire Service. 'My boss joined the Royal Army Service Corps (RASC) in July and then told his workers he wanted us to join up. Most went into the army but me and a mate didn't fancy that so we joined the fire brigade.' If Lew had thought the fire service offered him more chance of surviving to see his son grow up than the army he was wrong. 'I remember the start of the Blitz,' says Lew. 'On the way down to the Surrey Commercial Docks I looked up and saw hundreds of bombers coming over London. I turned to the bloke next to me and said "Christ, look at those bloody things".'

On 29 December Lew's station was called to fight a fire at the First Avenue Hotel, just north of Lincoln's Inn, between Gray's Inn Road and Warwick Court. First Avenue was a cut above its Holborn rivals – it was popular with businessmen who appreciated its modernity – but a ruinous fire had taken hold. 'We'd just started playing water on the hotel when the bomb dropped,' recalls Lew. 'They used to say that if you heard a bomb fall it wouldn't kill you; it was the ones you couldn't hear that killed you. Well, I heard this one and it's a sound I wouldn't wish on my worst enemy. The next thing I remembered was sitting on a bus with my wife on our way home to Willesden from the hospital – the blast had knocked me out and left me concussed but otherwise I was OK – and her saying to someone, "He looks a bit vague but he's quite all right!"'

Lew returned to Lincoln's Inn sub-station when he'd recovered. 'I asked one of my pals what had actually happened and he said "Lew, if you can't remember anything, it's best left that way".' At midday on 10 May Lew was doing the job he hated above others, 'washing our bloody appliances ... they always looked spotless to me but it was part of our duty'. With a bit of luck, however, Lew's duties would be finished by 4 p.m. and he had a couple of hours of Short Leave, enough time to tune into Forces Radio and catch Raymond Glendenning's commentary on the second half of the Cup Final from Wembley, Arsenal versus Preston. Short Leave was granted to firemen and women during a 48-hour shift. Some used it to sleep, some to shop, others to nip down the pub for a pint. 'We used to spend our Short Leave in the Clachan or one of the other pubs in the area,' says Lew. 'Spirits were rationed and the beer was weaker than usual but it was good fun. We were trusted to come back sober.'

4

LEW White's fire station at Lincoln's Inn was a five-minute walk down Chancery Lane to Fleet Street where Britain's newspapers had been based since the *Daily Courant* was published in 1702. Here egos and ambition jostled each other in the slew of Fleet Street pubs as journalists in their snap brim hats and buckskin shoes compared the largesse of their paper's expense accounts. Lew sometimes drank in Ye Olde Cock at the western end of Fleet Street opposite Chancery Lane. Nearby was El Vino's, favourite haunt of the journalists.

At the eastern end of Fleet Street, just before Ludgate Circus, there were two other renowned pubs, the Old Bell and The Punch Tavern with a model of Mr Punch sitting at the bar. The Old Bell, with its green and maroon façade, was where most of the *Daily Express* workers could be found propped up against the bar. Just up from the *Express* was the *Daily Telegraph*, a Union flag fluttering at the top of the vast white monolith, and the *Evening Standard* was at the back of the *Telegraph* in Shoe Lane.

In the centre of Fleet Street on the south side, situated on the corner of Bouverie Street facing Bouverie House, were the offices of *The Scotsman*, and Bouverie Street itself contained the *News Chronicle, Punch, The Economist, Star* and the *News of the World*. But it was the latest addition to Fleet Street, the Reuters Building at No. 85, opposite the *Daily Express*, whose influence reached furthest across the globe.

Edward Luytens had designed Reuters and, with its marble entrance and unequalled height, it was every bit as grand as the *Express* headquarters. Inside were housed several South African newspapers, the Associated Press and the Reuters news agency itself.

Shortly before midday on the morning of 10 May, John Hughes was heading to 85 Fleet Street to begin his shift at the Associated Press. He'd just come from the Lyons Corner House in Coventry Street where for his usual 1s 6d he'd polished off a breakfast of porridge, bacon and fried bread, dry toast and marmalade and a pot of tea.

As Hughes strolled down Fleet Street, past his fellow journalists, he did his best to conceal his contempt. It was only in his letters to his family back in Australia that he revealed his true feelings. 'The average Fleet St. journalist is quite the most flat-footed floogie, moron, imbecile and general nincompoop I have ever known,' he had told his daughter on one occasion. Hughes tried to steer clear of them, preferring to drink in the King Lud, a gloomy pub on Ludgate Hill famous for its Welsh rarebit.

Although he was an Australian, Hughes had spent much of his life in England. His first visit had been in 1909, as a member of the Australian rugby squad that toured Britain, and he was back a few years later as a member of the Australian Army. After the Armistice he settled in London, working first as a publicist and then as a journalist for the *Evening Standard*. By the time another war had broken out, Hughes was working for the Press Association. He sent his wife and daughter back to Australia, a severance that left him lonely and isolated. Gone were his daily confidantes with whom he used to express his astonishment of the British, a race he found easy to admire but hard to love. In weekly letters home, Hughes poured out his feelings. In March 1941 he had told his wife that the House of Commons was full of 'appalling Yes Men' who were trying to censor the Press, and he accused Winston Churchill of just 'reading his speeches'. A month later he reflected on the Blitz to his daughter Margaret: 'I honestly don't know how we stood it ... it was because the English are unimaginative and stolid. Nothing in the air, on the earth and on the water scares them. They get annoyed but they never get scared. I honestly believe that if 100 Nazi Paratroopers came down tonight in Piccadilly Circus they would start queuing up to see how our Commandos are dealing with them. The Londoner is great on queues. He queues for his bus, his breakfast, lunch and dinner, for the flicks and for the theatres.'

With no family to return to each night, Hughes often worked overtime to supplement his weekly wage of £11 1s 6d and during a slack period he could peer out of the window on the fifth floor of Reuters, east down Fleet Street, and see the top of St Paul's Cathedral. Beyond lay Aldgate, Whitechapel and Stepney. All these districts had burned fiercely on the night of 19 April, when the Luftwaffe did to the East End what they had done to the West End three nights earlier. One thousand tons of high explosive dropped and caused the deaths of an equal number of civilians. For Stepney it had been a night of destruction to rival the first few weeks of the Blitz when 40 per cent of its houses had been levelled. Most of Stepney's poor were crammed into cheaply made Victorian houses of thin brick and plaster, ready victims for the German bombs. A direct hit from a 50kg bomb would demolish one small brick-built terraced house or irreparably damage two; if a 250kg bomb hit the target it would flatten between four and eight terraced houses. But the East End's shoddy housing had its compensations. Rescue teams were able to tunnel through the wreckage and extricate people far more easily than with the solid masonry of the Georgian houses in Westminster and Chelsea.

Nineteen-year-old Florrie Jenkins had survived intact, as had her mum and her aunt's family. All of them lived together in a terraced house in Coutts Road, Stepney. 'We had three rooms and a kitchen upstairs,' recalls Jenkins, 'and three rooms and a kitchen downstairs with a toilet outside. My mum's sister lived upstairs with her children with me and my mum underneath. Every few years we changed over for a bit of variation.'

After leaving school Florrie had worked for the same company of wholesale grocers as her father in Whitechapel, she in the office as a clerk and he in the warehouse. Her dad had died in June 1939 so when the war came along, Florrie had been glad to leave behind the grocer's and its doleful connotations. She joined the Auxiliary Fire Service and was posted to Station 22 in Burdett Road, the next street up from her house in Coutts Road.

At just 5ft tall, Florrie was irrepressible; a bundle of fun, all Cockney banter and up-for-anything. 'During training we used to play cricket with the men and the pads were about as big as me!' In the pub across the road from the fire station Florrie played requests on the piano and drank and smoked with the firemen. She was known as 'Toots', because of her chatter, but Florrie was good at her job. 'I'd joined the fire service because I was interested in telephones and communications,' says Florrie, 'and when I'd been at school we had been taken on a tour of a telephone exchange and I really wanted to work in communications.'

During the raid of 19 April Florrie and the other three firewomen on duty had sat in the station control room logging reports of fires as they came down the telephones. They then passed these reports to the Mobilising Officer whose job it was to prioritize which crews attended which fires. On that particular night it had seemed as if Stepney was being engulfed by flames. And all the time Florrie had no idea if her mother was all right in Coutts Road. It didn't help her peace of mind that they didn't have an Anderson shelter out in the backyard. Like a lot of Londoners, Florrie's mum and aunt preferred to 'chance it' in the comfort of their own home.

There was nothing to stop an individual remaining in his or her house during an air raid; it was frowned on by the ARP wardens but they were powerless to force people into shelters. The most common shelter was the simple Anderson, just two curved sheets of corrugated steel fitted together and fastened to a rail. It was then sunk 3ft into the ground and sandbags were draped over the top and around the small entrance. A layer of soil was also sprinkled over the sandbags and over time people took to growing flowers or vegetables in this soil. The Anderson wasn't bombproof – a direct hit would obliterate it – but it was able to withstand the blast of a 250kg bomb from as near as 10ft.

What an Anderson shelter wasn't, however, was soundproof. If anything, its steel walls amplified the terrifying din of an air raid: the whistle of falling bombs, the cannon fire of the RAF fighters, the thunderous booming of the British anti-aircraft guns and, with a queer tingling sound, the shrapnel from those guns that fell on the people they were supposed to be protecting. There were other problems, too, oversights that shamed the Government. Andersons were built for the working class, a cheap but effective form of shelter. They were effective and they were cheap (families earning less than £250 a year qualified for a free Anderson shelter, otherwise they cost £6 14s) but they were really only for use in gardens, and 75 per cent of the population didn't have a garden in 1939, and London's working-class certainly didn't.

So in March 1940 the Government began constructing brick surface shelters, capable of holding fifty people from one street of a housing estate. They soon exhausted the country's supply of cement, but the Government ploughed on regardless, even without cement. They were complete by the time the Blitz started and the public soon learned they were cold and dark with desperate ventilation. With as many as fifty occupants in an airless brick shelter and just one bucket as a toilet in the corner, with a curtain to protect one's modesty, people entering a shelter gasped and spluttered as the stench hit them. 'We called them "splinter shelters",' recalls Joe Richardson, 'because they were useful if you were caught out in the open during a raid. You went inside to avoid the bomb splinters. But they were really just a nasty bodge job.' Others dubbed them 'Sandwich Shelters', a macabre nickname for a horrifying flaw in their design. If a bomb landed close by, the blast sucked out the walls and left the people inside the meat between the ground and the 9in slab of reinforced concrete roof.

For Londoners without an Anderson shelter and unwilling to risk being 'sandwiched', there were other alternatives. A hardy few roughed it in the trench shelters in parks and public gardens, and a meagre 4 per cent of Londoners took refuge in the London Tube. After much debate in Parliament, Churchill allowed supervised access to the Underground and many thousands headed down to the hot and dirty platforms. The majority didn't; the majority sneered at the Tube-dwellers.

In Stepney the most sought-after shelter was under the Tilbury railway arch. There was a labyrinth of cellars and vaults, able to accommodate 3,000 people, but as the bombing intensified, other areas were opened up and the capacity expanded to 14,000 on some nights. Sanitation was practically non-existent, and among the hundreds of families were prostitutes, racketeers, pickpockets and drunkards. Florrie Jenkins' mum preferred to remain within the four walls of her own home in Coutts Road.

At midday on 10 May Florrie was on duty at the Burdett Road fire station. Even though there hadn't been a major raid on London for three weeks, the control room always had to be staffed in case there was an ordinary factory fire in one of the many Jewish businesses dotted around Stepney.

THE quickest way from Stepney to south London was under the Thames through the Rotherhithe Tunnel. Halfway along the tunnel the tiles changed from white to black, and beyond this point Stepney's teenage gangs were in dangerous territory. 'The kids from Stepney weren't allowed past the white bricks,' remembers Tom Winter, who lived on the Redriff Estate in Rotherhithe.* 'The only time Bermondsey's gangs stopped fighting each other was when the kids from Stepney came through the tunnel. Then we joined up and fought them.'

When the Bermondsey gangs grew up they went to work either in the Surrey Commercial Docks or in one of the factories that throbbed and hummed and belched out pungent smoke; the leather tannery, Young's Glue factory, Hartley's Jam factory, Sarson's Vinegar distillery near Tower Bridge, Atkinson's Perfumery near Grange Road, and the Peek Freans Biscuit factory in Keeton's Road. For 12-year-old Tom the strongest smells were 'the hops that were stored in warehouses and the horse dung that was all over the road because horse and carts were still common'.

Tom's family had never had much money. When she was a child, Tom's mum was reduced to borrowing, begging or stealing fruit from the stalls at Borough Market, so hungry was the family. By the time war came round again, she was married with four sons and two daughters. Tom was the youngest son, wearing his brother's hand-me-downs and covering over the holes in his one pair of shoes with bits of cardboard. From an early age he used to watch his dad, Harry, earn a few extra quid in the boxing ring. 'He was a builder and quite handy with his fists so he'd go in the ring with the resident pugilist at The Ring in Blackfriars [a famous boxing venue] and any volunteer who survived three rounds got a few bob.'

Like a lot of children in London, Tom, two of his brothers and his

*Rotherhithe became part of the Borough of Bermondsey in 1900 and remained so until 1965 when it became part of the London Borough of Southwark.

two younger sisters were evacuated at the start of the war. 'We went down to Brighton but when the Germans invaded France my mum reckoned London was safer than the south coast so we came back.'*

In September they were bombed out of their house on the Holyoake Estate. 'So we went to live with our gran in Catford,' recalls Tom. 'After about four weeks we were bombed out of there, too.' The family was rehoused at No. 136 Redriff Estate. 'It was a ground floor flat with four rooms,' says Tom. 'Me and my two brothers slept in one bed and my sisters slept in a bed in my parents' room.'

There was no school for Tom, as that too had been bombed, so he and his mates amused themselves. Usually it was down by the river, which lay a couple of hundred yards away. 'One of the rackets,' remembers Tom, 'was fishing out lengths of good timber that had fallen into the Thames while being unloaded. We'd take it down to Customs and an official would measure it and give us a bob or two for our troubles.' And there were other prizes to be found in the river. 'We'd go up to Pier Head in Rotherhithe when the tide was out and collect incendiary bombs from the mud.'

A German incendiary bomb weighed about one kilogram. It was magnesium with a thermite filling, which could be removed by unscrewing the base of the bomb. One of Tom's gang, young Teddy Turner, fancied himself as a bit of a bomb expert. 'He would unscrew the cap,' recalls Tom, 'remove the phosphorous into a paper and then we'd all go and chuck it on a bonfire to see it go up.'

On Saturday 10 May there were no incendiaries for Tom Winter and his gang. The tide was approaching its highest mark, submerging the few incendiaries that so far had evaded the clutches of Bermondsey's teenage gangs. Instead of bomb hunting some boys played street cricket, chalking three stumps on to a wall and using a piece of wood

* By the end of August 1940, 2,500 children a week were returning to the capital; in early September 520,000 children of school age were in Greater London.

as a bat. Others split into two teams and played 'Hi Jimmy Knacker', a rougher version of leapfrog. A few hopscotched on paving stones fissured by bomb blasts. One or two chased the stray cats that lived in increasing numbers among the ruins, growing fat on the sewer rats that had been forced up into the open. Here and there, just visible in the spring sunshine, were the first pink flowers of the Rosebay willowherb, nicknamed 'fireweed' because of its predilection for growing in ground cleared by fire. It added a welcome dash of colour to Bermondsey's many bomb sites.

At 12.30 p.m. Tom Winter walked out of his flat and into another block on the Redriff Estate. On his right was the Fidgets pub, and 100yds on his left was the community centre where dances and wedding receptions were held from time to time. Tom entered another block on the estate and climbed the stairs to the top floor flat where his great pal, Ken Humphreys, lived with his mum and younger sister. The dad was away in the army, but Ken's paternal grandmother was in a ground floor flat 200yds away. As Tom and Ken went out to play in the bright sunshine they popped in to say hello to Granny Humphreys. She was a jovial old woman who never tired of answering the same question from Ken's disbelieving friends: yes, she was called Faith Humphreys and yes her middle names were Hope and Charity.

IN a few years Ken and Tom would graduate to the next stage of development in a Bermondsey boy's life: girls. Tony Donovan, Pat Leahy and Jimmy Gill, three 20-year-old mates now in the forces, had already reached the stage. Tony had fallen for Gladys Jenner, a brown-haired 19-year-old with an eye for the latest fashion and an inner toughness that was hidden by her demure exterior. 'On Sunday afternoons we girls donned our best clothes and paraded in the park at the bandstand for the boys,' recalls Gladys. 'That was where I met Tony.'

The pair lived close to each other and the courting was carried out in the dance halls of the New Cross Palais and the Hammersmith Palais. They talked about marriage and had discussed dates and venues, but both knew they were at the whim of the British Army. So while Tony trained with the Royal Engineers, the regiment he'd joined in May 1940, Gladys continued to earn £2 10s a week as a clerk in the offices of Peek Freans. In that first year of separation Gladys faced the greatest danger. Her family had been bombed out of their house in Keeton's Road in October 1940 so they moved into Clement's Road, just round the corner. Both houses were owned by Peek Freans and the Jenners were just one of many families who benefited from their company's munificence. But being a Peek Freans employee wasn't without its danger. All the tenancy houses were in the shadow of the enormous factory in Keeton's Road, and the Luftwaffe wanted it destroyed. It made ration packs and parcels for the British prisoners of war, and as a contributor to the war effort it was considered a legitimate target. 'It was well known that the POW parcels were made by us,' says Gladys. 'Firms like Cadburys and Bovril sent food, which was made into individual parcels and despatched to the Red Cross for distribution.'

At 1 p.m. on Saturday 10 May Gladys Jenner was walking out of Peek Freans. 'Weekday hours were 8.30 to 5,' she says, 'but on Saturdays we finished at 1.' Ahead lay a relaxing afternoon; perhaps a visit to her favourite dress shop in Peckham with a 'sign in the window saying "As worn by Judy Garland" or "As worn by Marlene Dietrich"', or maybe an afternoon dancing with her friends in Covent Garden's Royal Opera House. Since September 1939 it had been run as a Palais-de-Danse by Mecca Cafés, and Saturday afternoon was dance time. Two shillings for civilians, free to those in uniform.

At the Dockhead Fire Station in Wolseley Street half a mile north-west of Peek Freans, London's most celebrated firewoman was getting her hands dirty cleaning the chassis of her petrol wagon. And it was *her* wagon, even though she shared the driving with another female firewoman. It was the wagon that had earned Bobbie Tanner fame and a postbag full of fan mail.

Bobbie was 22 years old, robust in looks and personality, and born into immense privilege. Her maternal grandmother had married Charles Campbell, a colonel in the Blues and Royals and a good friend of King George V. She and her daughter had hunted with royalty near their house on Crown property in Northamptonshire. Bobbie's mother had married an adventurer who had spent time prospecting for gold in North America. When his daughter was born in 1919 he christened her Gillian Kluane, the two lakes in Canada he missed most, and within months of Gillian's birth his wanderlust overcame his parental responsibility. He never returned, and Gillian was raised by her mother who, at some point, decided to replace Gillian with Bobbie. They moved to Newmarket, so that Mrs Tanner could indulge her two passions in life: horse racing and gambling.

At 12, Bobbie was sent to France to learn the language. At 17, she was despatched to Germany, where her uncle was a naval attaché. Bobbie arrived during the Berlin Olympics in 1936 and was jostled by some Brown Shirts walking along the Unter Der Linden who disapproved of her lipstick.

She was back in England by 1937, living the life of the well-heeled country girl in Cirencester. She hunted and rode and looked after forty guinea pigs. Life was easy. The biggest challenge Bobbie faced was coaxing her front-wheel-drive car up the steep hills of Gloucestershire. When war was declared Bobbie drove to London and offered her services to the fire service. Perhaps it was an innocent posting or maybe the recruiting officer had a sense of humour; either way Bobbie was mortified when she saw her name beside StN61 Dockhead. She had expected Chelsea or Kensington, not Bermondsey. 'You can't send me to a place like that!'. The officer looked at her and smiled. 'See how you get on. If you don't like it after a couple of weeks we'll see if we can transfer you.'

Dockhead Station was set back a couple of hundred yards from the banks of the Thames. There were docks to the east and across the river in Wapping. Tower Bridge was a short walk west. Few fire stations in London were as exposed as Dockhead and from the day the Blitz started Bobbie was in the eye of the storm. As one of the few women drivers in the fire service, her skills were put to most effective use; she drove the 30cwt wagon lorry, filling up the trailer pumps as the incendiaries fizzed all around. It required a steady hand and boundless courage, two qualities Bobbie possessed in abundance. In late 1940 her persistent valour was rewarded with the announcement that she had been recommended for the George Medal. As the first firewoman to be honoured with the award, the press and public alike went to town. Newspapers fished for an angle, something with which to reassure their readers that Bobbie was a really just a girl-next-door type. They asked her if she knitted. She didn't, she preferred to hunt and shoot. The papers ran the story anyway, proudly telling readers that after raids Bobbie found knitting the most soothing way to unwind.

Bobbie laughed it off but the Women's AFS newsletter of February 1941 was not impressed with the coverage. It congratulated Bobbie on her medal, 'though to read some newspaper accounts one would think it was almost equally due to her capacity for knitting without dropping stitches during a raid'.

Bobbie received the medal from King George VI in February 1941 at Buckingham Palace. The King pinned the George Medal on her tunic, then offered a few words of gratitude. He stammered and stammered, so Bobbie finished the sentence for him.

At 2 p.m. in Peckham, just south of Bermondsey, 30-year-old Gladys Shaw was setting out on her Roadster bicycle to ride the 9 miles to Keston fishponds in Kent. Her pockets bulged with a flask of tea and sandwiches wrapped in greaseproof paper. A small, old-fashioned young woman with a keen cheery face and a quiet calm voice, Gladys was looking forward to seeing some green.

Kent was where Gladys had been born in 1911, the daughter of a journalist who got a job with the *Sussex County Herald* and moved the family to Brighton in the 1920s. He was horrified when Gladys had told him she wanted to be a missionary. Religion had never played a part in the Shaw household. What he'd seen in France in the First World War had convinced Mr Shaw there was no God. But his daughter was adamant. In 1936 she caught the train to London and joined the Ranyard Mission.

Gladys learned on the job, visiting the most hopeless slums of south London with the vicar of St Mary's Church in Peckham. At the end of her first day she went back to her digs and wept, crushed by the misery in twentieth-century London. In the weeks that followed, Gladys realized the extent of her sheltered upbringing. Illegitimacy and promiscuity were rampant; their only equals, poverty and despair.

It took Gladys six months to be accepted by the people of Peckham. When they invited her in for a cup of tea she knew she had won their trust. When war broke out there was no question of her leaving and heading south to her family on the coast. She would stay and suffer with her adopted family.

As well as continuing with her missionary work, Gladys volunteered to work two nights a week at a first aid post and two nights a week as a firewatcher. When the bombs started falling she visited street shelters and read passages from the Bible to comfort the scared. But Gladys was as scared as everyone else. That was why she pedalled out of the city whenever she could. There was still beauty in the world but you had to get out of London to discover it.

5

THE men of 29 Squadron were fond of West Malling airfield. It catered for their every whim. London was within easy travelling distance, the pubs were plentiful and the fields were full of Land Army girls. This was Kentish hop country and the aroma that drifted lazily through the May sunshine put them in mind of the carefree days of summers past. To the men the dark cordon of trees that hugged the airfield's perimeter were symbolic; they had been there for centuries and like England herself they would remain, tall, upright and immovable.

It was just after lunch on 10 May, eleven days since 29 Squadron had arrived from their old base in Lincolnshire. The quarters were in the Old Maidstone Flying Club clubhouse. On one of the walls hung a picture of Amy Johnson, who had flown here in the 1930s. The Officers' Mess was in The Hermitage, a Victorian country house on the other side of West Malling village. It was as austere and aloof as the century in which it was built. The grand oak staircase that led from the entrance hall to the officers' rooms on the first floor creaked and groaned as if in protest at the intrusion of the young upstarts.

The sergeants were billeted in Malling Abbey, built in the eleventh century and reputed to have been a hiding place for the assassins of Thomas à Becket during their flight from Canterbury. That was one of the tales told to the sergeants of 29 Squadron when they drank in the *Startled Saint* pub. Another one was more recent; it concerned the Spitfire pilot who had been stationed at West Malling during the Battle of Britain. After he'd nursed his perforated aircraft down safely he was stopped on his way to the watch office by one of the labourers tempted down from the north to strengthen the airfield's defences and earn the sort or money that was scarce back home. The labourer offered the young pilot a cigarette and the two sat on the grass smoking. 'How much do you get paid?' the labourer enquired.

'About six pounds a week.' The pilot took a deep drag on his cigarette. 'What about you?' he asked.

'Eight pounds,' the labourer told him. 'But of course I work through air raid alerts.'

The workmen had returned north now, and their place in the pub was taken by the Land Army girls. They held no interest for Sergeant Richard James, one of 29 Squadron's most experienced aircrew and one of the few married men. He was a Lancastrian of substance, a clean-limbed steady sort. Before the war he'd managed the paints department of a builders' merchants. Now he was a radar operator in a Beaufighter nightfighter, a pioneer in a new form of aerial warfare.

His pilot was a 24-year-old flight-lieutenant with a DFC. He was short and stocky and talked with the waspishness of inferiority. 'Guy Gibson joined our squadron at the end of 1940,' recalls James, 'and we all thought he looked too young to be a flight-lieutenant with a DFC.'

Gibson had been posted to 29 Squadron from Bomber Command for a rest. His nerves, as he himself admitted, were 'beginning to get affected' and few were the nights when he wasn't tortured by nightmares. Of the pre-war members of 83 Squadron, Gibson was the sole survivor after six months of bombing missions over France and the Ruhr in 1940.

When Gibson had first walked into 29 Squadron's mess he entered a different world. 'They were all dressed in the conventional attire of the fighter boy,' he wrote, 'flying boots with a couple of maps stuck in the top, roll-neck pullovers and/or sweaters, no ties or collars and, of course, dirty tunics with top buttons undone. I felt a little embarrassed standing there looking like Little Lord Fauntleroy ... it was obvious that these boys weren't pleased to see me.'

Gibson was right. His arrival was resented, and not just because he was taking over command of 'A' flight at the expense of a popular officer. The DFC ribbon on his flying tunic jarred with pilots who were thoroughly browned off with their lot. Since the start of the war James and 29 Squadron had been chasing German shadows in their Blenheims. 'The Blenheim was a horrible aircraft,' says James. 'And of course it wasn't as fast as the planes we were supposed to shoot down.'

Gibson's posting to 29 Squadron coincided with the conversion from Blenheims to Beaufighters. The Beaufighter, wrote Gibson, 'was the new plane of the day', and great things were expected from it when it came into service in September 1940. Its maximum speed of 323mph at 15,000ft ridiculed the Blenheim's best, and its armament of four 20mm Hispano cannon and six .303in machine guns aroused envy among Spitfire pilots. But RAF Fighter Command was most excited by its radar, the very latest with a higher powered transmitter and a receiver that made it possible to track targets to a minimum range of 100yds. The Blenheims had carried gunners but these were superfluous in the Beaufighter, whose armaments were operated by the pilot. When the Beaufighters were delivered to 29 Squadron, air gunners like James knew they 'would be out of a job unless we trained on AI [Airborne Interception]'.

James prowled warily round the Beaufighter, looking her up and down. Then he climbed inside, through the panel in the bottom of the fuselage that hinged downwards. Once inside it was a squeeze. Right in front, halfway along the fuselage, was a swivel-seat with a back rest for the radar operator. From the seat he could look out through the perspex dome. There was a purr of approval from James; it gave an unobstructed view and with a little wriggle he could see below and behind. That had been the Blenheim's blind spot for an air gunner.

The AI radar box was suspended from the low roof just behind the dome. It was a queer-looking thing with a rubber visor and a set of control knobs. Instruction in its mysterious ways would come later; first James wanted to examine the rest of the aircraft. From the radar operator's seat, he shuffled his way along the catwalk towards the armour-plated doors behind which was the cockpit. Either side of the catwalk, set in racks, were eight ammunition drums for the cannons that were just below floor level. James felt the weight of one. 'Very heavy,' he recalls. 'More than 60lbs each.'

For a month either side of Christmas 1940 the aircrew of 29 Squadron acquainted themselves with their new aircraft. Gibson found it 'possessed a terrific punch in its armament ... [but] for its size the Beau was very heavy and its wing loading rather on the high side. For this reason the CO insisted on all pilots carrying out a lot of flying before trying to fly it at night'.

James's keen brain had grasped the concept of airborne radar more quickly than most and he was asked to 'train the half a dozen radar operators' posted to their base in Digby, Lincolnshire. For the uninitiated the first look into the leather visor was confusing. When the AI was switched on, the two cathode tubes, one horizontal and the other vertical, became luminous green lines. James explained that it was really very simple: the AI emitted a series of radio blips. If there was an aircraft within range an echo bounced back. It appeared on the green lines as a cluster of sparkling lights. 'These were known as blips,' explains James, 'and if the aircraft climbed higher the blip [on the vertical tube] got bigger so the radar operator told the pilot to climb.

From the blip on the vertical tube you could see if the aircraft was port or starboard.' The radio operator gave instructions to the pilot – increase throttle, decrease speed – until the blips sat squarely across the tubes. Then the aircraft was visible to the pilot, and he chose the moment to attack.

The weeks allowed 29 Squadron to adapt to their new aircraft were appreciated but frustrating none the less. And while they trained, German bombers attacked Britain with impunity. London. Coventry. Hull. Liverpool. Portsmouth. The civilians wanted to know why more wasn't being done to protect them. With the firemen now 'Heroes with Grimy Faces', some of the people sought out new victims to harangue. Gibson was in London on a spot of leave with his new wife when the air raid siren started to wail. They took shelter in an Underground station. 'As I walked along the rows upon rows of huddled figures crouching on each side,' he said, 'I was aware of a certain animosity in the crowd. Suddenly one woman yelled at the top of her voice, "Why don't you get up there and fight those bastards?" I quickly turned and ran up the stairs again, not wishing to be torn limb from limb.'

There was further frustration in January 1941 for 29 Squadron. Thick fog loitered on the ground for three weeks and crews managed only one sortie. By the time the fog had cleared in February, Gibson had chosen his radar operator. 'He came up to me,' recalls James, 'and asked if I would like to fly with him. I'd never flown with Gibson before but I think he thought I was one of the best as I'd done a lot of radar instruction.'

James was under no obligation to accept the invitation. He had flown a couple of sorties with another officer whom he liked, and Gibson's reputation preceded him. To the ground crew of 83 Squadron, he had been known as the 'bumptious bastard' and his aloofness hadn't endeared him to all of 29 Squadron. He didn't talk to his ground crew, the men who ensured his aircraft was always airworthy, because Gibson didn't really talk to anyone other than fellow officers.

But Gibson's brittle personality didn't bother James. 'When he asked me I thought two things,' he says. 'That he must think I'm good and that I'll live longer with this chap.' At this stage of the war, James wasn't concerned with survival, just with living for as long as possible. 'Some airmen had an attitude of "It will never happen to me". That wasn't my philosophy. I knew it jolly well could happen to me because I had seen so many people lost to enemy action or bad weather or aircraft problems. But as I'd made up my mind that I would die I wasn't worried about dying.'

Their first patrol together was on 1 February 1941. Three days later James picked up an enemy aircraft on the radar and vectored (steered) Gibson to within sight of it. He opened fire but the German escaped.

Over the next six weeks their experience and understanding of one another took root in the grey and cheerless skies of eastern England. 'Flying with Gibson was terrific,' says James, 'because he was a first-class pilot. He could land a Beaufighter better than anyone in the squadron, even at night time … he also practised landing on one engine at night.' Even when they took off in atrocious weather, James had every confidence in his pilot. 'There would be literally no horizon but he was a very good instrument flyer and he used to tell me "you've got to be able to take off entirely on instruments in this job".'

Gibson's respect for James's skills grew as February turned to March, though one wouldn't have known it. 'When I'd first started flying with him I couldn't get through to him. He called me James, while everyone else used my nickname, "Jimmy". He never asked about my family or me. I couldn't understand it because all the other crews were on first name terms regardless of rank.' At least Gibson occasionally engaged James in conversation. His ground crew were ignored. 'After a flight he would walk past his ground crew without saying a word,' says James. 'At first I did the same but then I thought "bugger this" and stopped for a chat. They always wanted to know if we'd been in action and what had happened.'

James says that because of youth and his size, Gibson had a 'inferiority complex'. Yet by now he was respected by all in 29 Squadron. 'I was in absolute awe of him, and quite frankly he scared the shit out of me, with the things he did.'

On one night patrol over the Channel, James pointed out a 'massive searchlight off Dunkirk that appeared to be lighting up the English coast'. 'He said over the intercom, "Yeah, it is big, isn't it? Shall we put it out?" What could I say? If I'd said no he would've thought I was chicken. So I said "Yeah, go on then". We dived and nothing happened until we were right on top of the searchlight. Then we opened up with our four cannon and machine guns and put it out. But the stuff that came back at us ... oh, it was grim. We very wisely made a short turn to port and kept as low as possible so nothing could hit us.'

By mid March the Germans had intensified the Blitz, as if to compensate for the lull caused by the previous bad weather. Cities and ports on the east coast were targeted on 14 March, and 29 Squadron were patrolling when James picked up a contact on the radar. He guided Gibson on to the tail of a Heinkel He111 but when he pressed the firing button nothing happened. Twice more that night Gibson manoeuvred into position to open fire but each time the cannon jammed again after a short burst. Gibson yelled to James through the intercom. 'I had to clear [the blockage] in the cannons,' remembers James. He bumped along the catwalk and began to change the ammunition drums. There was an intercom plug nearby so he could keep in contact with Gibson, but 'they hadn't bothered to fit any oxygen amidships'. James recalls that with each drum weighing more than 60lbs, it was 'bloody hard work and just after I cleared the last gun

I collapsed. But with that gun we shot it down. Gibson couldn't hear anything for a while until I came to'.

Other crews enjoyed similar success that night and the squadron threw a party to celebrate. Gibson stood James a couple of beers and addressed him as 'Jimmy'.

The next month they 'intercepted and damaged' a Dornier, and on 3 May, four days after the move to West Malling, they destroyed a German bomber over Southampton.

For the next week Gibson and James patrolled the skies over southern England; two hours on Monday night, three on Wednesday evening, and two hours forty-five minutes on the evening of the Friday. No contact was made with enemy aircraft.

In the early afternoon of the Saturday, 10 May, Richard James had a few hours to himself. Friday evening's patrol had lasted until 0110 hours and he knew he would be up again tonight. Gibson was visiting his wife who lived in a nearby cottage, and the usual afternoon AI practice had been cancelled. It was a pleasant afternoon to have off; the sun had warmed the woodland and hop fields, though there was still a nip in the air. The West Malling cricketers limbering up for that afternoon's match wore sweaters, and the shoppers pottering down the village's high street, lined with Tudor, Jacobean and Georgian houses, didn't loiter too long chatting to each other.

Suddenly the rural tranquillity erupted in a deafening cacophony. One, two, three, four, five Spitfires roared overhead, unmistakable with their elliptical wings. The sweet purr of their Merlin engines brought villagers running excitedly out of their houses. More Spitfires followed, swooping over their houses and heading towards the airfield in a curved approach.

6

PILOT Officer Freddie Sutton of 264 Squadron watched the undercarriages of the Spitfires dropping as they came in low over the trees. One or two came in rather fast, and for a moment seemed to float, before the wheels touched the ground. Others, he noticed, the more experienced pilots probably, kept the nose well up to control engine power and with good positive use of the throttle landed in a very short space. The pilots of 264 Squadron swapped admiring glances, none of them feeling any pique at the sight of a squadron of Spitfires landing on their airfield. 'For some unknown reason they [the day boys] had a tremendous admiration for the nightfighters,' said Sutton. 'This mutual respect naturally made for a friendliness from the start.'

Sutton had applied for a commission in the RAF as an air gunner days after the declaration of war. At his selection board one of the officers had told him that after three months in uniform he was unlikely to be alive. 'It struck me at the time as being a rather queer method of attracting recruits,' he wrote in his diary. Sutton had grinned boyishly, as he often did. With his long face and ready smile it was hard to take offence at Sutton. 'I just told them that I knew nothing about guns and have never even seen an aircraft on the ground.'

He was accepted for training, however, and in July 1940 arrived at 264 Squadron. His fellow gunners were a disparate bunch, one 'was an ex-member of a Glasgow Street corner gang [and] in between we had schoolmasters, actors, stunt men – just a cross-section of life'. The squadron flew Defiants, two-seater fighters with a pilot and a turret gunner. He had four .303 Browning machine guns that could traverse through 360° and elevate from horizontal to vertical. The Defiant had been intended for offensive patrolling or as a bomber destroyer but the Battle of Britain bled the RAF practically dry of aircraft and the Defiant was pressed into service. The German fighters found them easy meat. On 28 August, a month after Sutton's arrival, 264 Squadron lost four aircraft to Messerschmitt 109s. Three more were damaged and their aircrews 'took silk' as the RAF called baling out. That in itself was a struggle in a Defiant. There had been cases of gunners unable to escape because they couldn't squeeze out of their cramped turrets.

But life had improved since then. Lessons had been learned and 264 Squadron was used as a nightfighter squadron when the Blitz started. The Defiants didn't come equipped with radar. They were a 'Cat's Eyes' squadron who relied on sharp eyesight alone. It was, reflected Sutton, often a 'life of disappointment ... chasing shadows and never seeing anything'. But on 8 April Sutton and his pilot, 264's commanding officer, Squadron Leader Arthur 'Scruffy' Sanders, shot down a Heinkel. Back in his quarters, his pen dipped in adrenaline, Sutton told his diary of the encounter: 'My four guns roared into life as they spat death at the rate of 4,800 slugs a minute.'

Six days later 264 Squadron moved to West Malling. At first they had been the sole occupants of the Kent airfield, then 29 Squadron arrived a fortnight later. Sutton was able to brief the gatecrashers on the charms of their temporary home. 'The trains took exactly 50 minutes to Victoria and there was one every hour of the day. The nearest town was but five miles away and the district abounded with the finest of old country pubs [and] the station had been rebuilt entirely for the accommodation of nightfighter squadrons.'

THE Spitfires taxied to a stop and pilots started emerging from the cockpit. They jumped down on to the grass and moved towards the clubhouse with the nonchalant maunder of men with nothing to prove. The crews of 29 and 264 Squadrons smiled to themselves as the pilots of 74 Squadron – 'Tiger' Squadron – approached. Those among them who were aware of their RAF history knew that this was the squadron of the legendary Mick Mannock, Britain's highest-scoring First World War ace.

The squadron had been flying Spitfires since February 1939, the second RAF squadron to be equipped with the new aircraft. In July that year they had been the guests of honour in Le Bourget to celebrate Bastille Day and the two German officials, representing Lufthansa, had taken a keen interest in the aircraft while the pilots were presented to a French general, who moved among the young men, commenting on the cosmopolitan composition of the squadron. The CO was Squadron Leader George Sampson, an Australian martinet; there were two Irishmen, Paddy Byrne and Paddy Treacy, who had been to the same Dublin school and disliked each other with a Celtic passion; an arrogant-looking South African called Sailor Malan; and happy-go-lucky New Zealander, Don Cobden, who had played rugby for the All Blacks. There were even a few Englishman: H. M. Stephen, Gordon Heywood, Peter Chesters and 19-year-old John Freeborn. The French General stood before Freeborn and gave him a paternal pat on the cheek. 'So you have brought your mascot then?' he said, turning to Sampson. Freeborn, electrically charged with indignation, bit his lip and looked straight ahead. Sampson breathed a sigh of relief at the teenager's restraint.

Pilot Officer Freeborn was the 'pup' of the squadron. He was short and unremarkable-looking, with a fascination for steam locomotives. With a discreet moustache and brown hair sensibly parted, Freeborn didn't look like an ace fighter pilot. But in the air he was a killer, and only Malan could match his skill. Freeborn knew how good he was; his self-confidence, never mottled with arrogance, allied to his Yorkshire temperament, had led him into trouble down the years. At Leeds Grammar School Freeborn had beaten up a teacher after he'd had a ruler broken over his head. At 16 he had left school, at 17 he joined the RAF and at 18 he was posted to 74 Squadron. 'I arrived on a Sunday night,' he recalls, 'and the orderly officer fixed me up with a room, bought me some beer and sent me to bed cross-eyed.'

The next day Freeborn was up in the air in a Gloster Gauntlet, an obsolescent biplane with a top speed of 230mph. It was the squadron's aircraft until they were re-equipped. When the Spitfires arrived, the pilots of 74 Squadron each felt like a man on a date with a beautiful woman; wanting to touch but overawed. Freeborn had never before flown a monoplane and his first thought was 'she was rather nose heavy'. Paddy Treacy sat him in the cockpit and went through the rudiments. Then Freeborn taxied down the runway, the long nose of the Spitfire hindering his forward vision as he gathered speed. 'It went so bloody fast,' he recalls, 'that I shot between two hangars at 180mph before I took off ... once at altitude I settled down and found the Spitfire to be a lovely aeroplane. You became wholly part of it.'

Six weeks after the Bastille celebrations in France, 74 Squadron was at war. They went into action for the first time on 4 September, but it was a false alarm. Two days later, at 0700 hours, they were scrambled again from their base in Hornchurch. Red Section of 'A' Flight led the way under Malan's command. Yellow Section climbed after them, Paddy Byrne with Freeborn and 'Polly' Flinders as his Nos. 2 and 3. Two aircraft were spotted over Barking Creek and Malan gave the call over his intercom. 'Tally Ho! Number one attack – Go!' Byrne and Freeborn dived on the aircraft and opened fire, 'pressing the tit' they called it. They fired their eight .303 Browning machine guns in rapid two-second bursts, as they'd been trained to do.

The two aircraft caught fire and began their steep uncontrolled dive towards the Essex countryside. One baled out, but even before the other slammed into the ground the pilots of 74 Squadron knew they'd made an appalling mistake. The aircraft were Hurricanes belonging to 56 Squadron.

'When we landed, Byrne and I were placed under close arrest,' says Freeborn. The hearing took place a month later with Malan appearing for the prosecution and insisting he'd ordered them to break off the attack. 'He dropped us right in the shit,' Freeborn says. 'It was Malan's bloody fault. He gave the order to attack and we attacked. There was no order to break off.' 'Polly' Flinders agreed with Freeborn and the pair were acquitted. 'I was very sorry about it,' says Freeborn.

Freeborn shot down his first German the following May over Dunkirk. It left him feeling 'bloody good'. 'I turned into the clouds and came out dead behind this German. It was a Messerschmitt 109. I opened fire and he went down. One wing caught a telegraph pole and he smashed into a little French farmhouse.' The farmer, ploughing his field, looked up at Freeborn and made clear his fury with some Gallic gestures.

By the time the Battle of Britain started a couple of months later, Freeborn had shot down several more Germans. Some of his colleagues disapproved of his methods, but to Freeborn 'they were bloody Germans and they had to go'. A fellow pilot, Bill Skinner, recalled a Ju88 that had been shot down by 74 Squadron. 'It crashed through a herd of cows, tossing some of them in the air. The Nazi crew were seen to get out of their machine and proceed to shake their fists at us. That proved too much for Freeborn. He promptly shot them up and consigned them to their Maker.'

Freeborn also liked to put the frighteners on any German who baled out. 'My view was that it was fine to give them a scare,' he says. 'I had an argument with Stevens about this and I used to take the piss out of him, saying "You should have seen the one today, he was climbing up his parachute lines!" I never shot one in his parachute, but I flew close enough to scare them.'

Freeborn was easily identifiable as he strolled across West Malling airfield at 2.30 p.m. on Saturday 10 May. He was shorter than the rest, most of who were clean-shaven. Freeborn had grown a neat little moustache, but if he'd hoped it would add on a couple of years to his looks he was mistaken. He looked his age, 21, despite having come through eighteen months of intensive combat. Of the pre-war originals, those pilots that had taken part in the Bastille Day celebrations two years prior, Chesters, Cobden and Byrne were dead. The rest were either prisoners of war or, like Malan, serving with other squadrons. But Freeborn was still there, as dogmatic and pugnacious as ever, but now with a DFC and bar ribbon on his tunic, 12½ kills to his name, and having flown more operational hours during the Battle of Britain than any other pilot of any other squadron. The newer pilots to Tiger Squadron had come to regard Freeborn as indestructible. Others came and went – three pilots had been shot down four days earlier during a bomber escort over France – but Freeborn flew on. 'We coped with the regular losses,' says Freeborn, 'by becoming callous and indifferent.'

There were still a handful of Battle of Britain veterans in the squadron, pilots such as the perpetually smiling Peter Stevenson, the debonair Tony Mould and the chain-smoking Roger Boulding. Much to Freeborn's delight, Boulding's pet dog, Sam, a brown and white cocker spaniel that used to interfere with squadron card games, had been left in Gravesend. If Freeborn had had his way, 'we would have eaten the little bugger!'

Boulding looked as though he should have been playing trumpet in a London nightclub band; there was a cigarette permanently dangling over his lower lip and his brown hair was cut fashionably short, then greased back with Brylcreem. But he liked to have a few strands falling rakishly down above his right eye.

Boulding had joined 74 Squadron in August 1940, 'having been with the Advanced Air Striking Force in France flying Fairey Battles'. After the Battle of Britain, Boulding had flown as Malan's No. 2 on some of the sweeps across the Channel. These sweeps, Boulding said, were carried out in the hope that 74 Squadron could 'persuade the enemy to send up fighters so we would have an opportunity to knock them down'. More often than not, the Germans refused to play the game and with the bad weather of early 1941 Boulding hadn't shot down a single enemy aircraft by the time the squadron arrived at West Malling. He had damaged a Messerschmitt 109 on 7 May over North Foreland – he'd seen pieces of the plane break off – but as Freeborn teased him that didn't count as a 'kill'.

As he approached the West Malling clubhouse, Boulding knew that he might soon have the chance to break his duck. Tonight 74 Squadron would be 'involved in reinforcing the night defences of London during the three days of the full moon period'. He and Freeborn bantered about what they wanted as souvenirs if either shot down an enemy bomber. Boulding was after a dingy, for use on the lake at Cobham Hall where they were billeted in Gravesend; Freeborn 'wanted one of the 9mm Lugers that were carried by the German crew'.

7

THE countryside around Vannes in Brittany was as lush as its Kent counterpart across the Channel. On this early summer afternoon the green and blue and yellow gave the crews of Kampfgruppe 100 (KGr100) a sense of peaceful contentment. Like West Malling, the airfield at Vannes/Meucon had no hard-surface runway. Nor did it have proper quarters such as those used by the crews of 29 and 264 Squadrons. The only buildings at Vannes were one small open-ended hangar, an old wooden barn converted into a workshop and a cluster of wooden huts that served as the Operations Room HQ. KGr100's commanding officer, 32-year-old Hauptmann Kurd Aschenbrenner, considered it unworthy of his squadron and fit only for what it had been before the war: a small aerodrome for private flying enthusiasts. None the less this is where they had been stationed and they had to get on with it. And there were compensations; the crews of KGr100 were stationed 3 kilometres east of the airfield, in Vannes' Hotel le Bretagne, and the nearby beaches were ideal for frolicking with local girls. When crews had completed a specific number of missions over Britain they were sent slightly further afield, to the Hotel Boris on the Port Navalo coast, which had been designated a special 'rest and recreation centre' for exhausted air crews.

KGr100 were the Luftwaffe Pathfinders and to the crews they were the best pilots Germany had. In November 1939, having taken part in the defeat of Poland, KGr100 was granted full status as an independent Bomber Wing, and six months later their Heinkel He111s were decorated with the unit's new emblem, a Viking ship with a black hull and a red and white sail. In April and May KGr100 had taken part in the Norwegian campaign, but in June they were withdrawn to Luneburg, Germany, to familiarize themselves with the latest navigational system.

The British were aware Germany had such a system, but that was all they knew. In the same month that KGr100 began training on the new method, Churchill was informed that the Luftwaffe was 'preparing a device by means of which they would be able to bomb by day or night whatever the weather. It now appeared the Germans had developed a radio beam which, like an invisible searchlight, would guide bombers with considerable precision to their target'.

The device Churchill had been warned about was *'Knickebein'* (bent leg), but KGr100 had been chosen to pioneer the latest system, called 'X-Gerat'. It worked on the principle, first developed by Germany in the 1930s for use in civil aviation, of using radio beams for aircraft directional guidance. One pilot wrote in his diary on 18 June 1940 that they were all pleased with X-Gerat. 'I am pretty sure that when it is in use it will be a success, for today when we flew on an X-system training flight it produced excellent results.' All they were waiting for now, he added, was the installation of the transmitters, so they 'can use them on our forthcoming operations'.

By the end of July a chain of radio transmitters had been erected on the northern French coast. In the first week of August KGr100 moved to Vannes, ready to begin attacking Britain.

They coated the undersurfaces of their Heinkels with matt black distemper, then did the same to the fuselage to reduce still further the chances of being caught in a British searchlight. On 14 August, twenty-one aircraft from KGr100 took off from Vannes in their inaugural attack on Britain. The target had been some factories in the Midlands. In their briefings KGr100 had been told that the radio beams of two of the radio transmitters would intersect over their allotted targets. The rest was down to the crews once they were airborne.

When they were in the air the radio operators (r/o) switched on their Lorenz receivers and tuned in to the frequency between 30.0 and 33.3kHz. The steady monotone signal in their headsets told them they were following the radio beam. If the r/o heard dots or dashes the pilot had strayed away the beam and was off course. The bombers flew down the beam until the r/o heard the 'advance signal' on a second receiver. This was another beam laid across the main one they were following. The advance signal told the crew they were 12 miles from the target. Six miles further on they crossed a second beam and the r/o pressed a button that started the time clock. When they cut across a third beam the r/o pressed the time clock again. One of the clock hands stopped. The second hand continued until it reached the first and the electric contact released the bombs over the target.

When the Luftwaffe began bombing London on 7 September, KGr100 had targeted two gas holders between Chelsea and Vauxhall Bridge. But within weeks it was decided to use the unit as Beleuchtergruppe, or 'fire-lighters' for the main bomber force behind them. Throughout the autumn of 1940 KGr100 led the way in the Blitz over Britain, dropping mainly incendiary bombs to illuminate the targets for those that followed. In theory it was a hazardous task, but KGr100 had grown nearly as contemptuous of the RAF fighters as they had of the inaccurate British Ack-Ack guns. 'During our return flights,' wrote one, 'the radio operator would often tune in his receiver to a music programme to provide some relief from the monotony.'

But by the early spring of 1941, KGr100 had become ever more unsettled by the success of the RAF nightfighters. They knew that the British had got wind of X-Gerat, and were doing their best to interfere with the radio frequency, but they weren't sure how the fighters were able to find them in the inky blackness of 15,000ft. They knew nothing about airborne radar, so they assumed the fighters flew out on the same navigational beam that was supposed to guide them to their target.

As for the stories that had started to appear in the British press under such as headlines as 'Why Carrots Help You To See in the Dark', they weren't sure what to make of them so they put it down to British eccentricity, even when one of the RAF's top pilots, John 'Cat's Eyes' Cunningham, began to extol the virtues of eating carrots. 'What a load of rubbish that all was,' recalls Richard James, never a carrot fan. 'Still, it gave us all a good laugh when we read about it.'

The new-found respect for the British nightfighters in the spring of 1941 was based on the grim tally of the last four weeks. Six aircraft of KGr100 had been destroyed or damaged in bombing missions over Britain. In the first three months of 1941 they had lost only two aircraft. Radio operators had stopped tuning in to the latest jazz numbers on their return flights.

The crews of KGr100 had no idea if they would be flying on the night of 10 May. They suspected they would, though they didn't know where. Three nights ago it had been Liverpool, the next evening Hull and Nottingham. With perfect flying conditions forecast for the evening the air crews knew they would be ordered into the air. Where to would be revealed when they were called to the Operations Room for an early evening briefing.

Elsewhere across northern France, German bomber crews lazed in the afternoon sun waiting for their orders. At the Chateau Roland, near Versailles, the crews of Kampfgruppe 55 chatted in the sprawling grounds or kicked a football around. At Vendeville, near Lille, 25-year-old Hauptmann Albert Hufenreuter galloped across the monotonous countryside on a horse he'd borrowed for a couple of hours. He preferred his native landscape, the thick fir forests and dramatic beauty of the Harz Mountains in Lower Saxony in which nestled his home town of Quedlinburg. With its labyrinth of cobbled streets and half-timbered medieval houses, Quedlinburg looked like a setting from a Brothers Grimm fairy tale. But life had been tough in the years after the First World War for the Hufenreuter family. Albert had dreamed of becoming a farmer, but his father's paltry teacher's salary wasn't enough to pay for a college education, so instead he joined the air force. His broad shoulders and dark looks filled the uniform well, and his confident intelligence caught the eye of his superiors. When war broke out Albert was working in the Air Ministry, and it was only after persistent badgering that he achieved his wish to join an operational bomber squadron, by which time he'd taught himself the skills of night navigation.

Hufenreuter had now flown twenty missions over Britain but on the afternoon of 10 May 1941 the furthest thing from his mind was the twenty-first. As he cantered homewards, Hufenreuter glanced at his wristwatch. It was 3 p.m.

8

THE clock in the Wembley dressing room showed ten minutes past three. With every second that ticked by the hum of the crowd increased. The hubbub sent butterflys fluttering round the stomach of 19-year-old Tom Finney as he laced up his boots alongside his Preston North End team mates. This was his first taste of the big time, a step up from the league games for which he got paid £2.10s a time. His brother Joe was one of the 60,000 spectators, one ripple in a sea of khaki uniforms that had come to see Preston fight Arsenal for the Cup.

Tom didn't give a hoot that a week earlier Emanuel Shinwell, a firebrand Labour MP, had condemned the continuance of football at a May Day parade. 'I'm no killjoy,' he bellowed. 'I want to see the people of this country getting a fair measure of recreation but when I hear of 60,000 gathered at a football match I wonder whether we are crazy … think of the petrol consumed, the transport used and the services required for all the so-called recreation and ask yourselves if we're really organising our resources for war.'

Shinwell's was a lone voice in a country hungry for some light relief from the gloomy war news. On the Monday after Shinwell's tirade, 5 May, it was announced that the Cup Final between Arsenal and Preston North End would be all-ticket. By the end of the day all 40,000 standing room tickets had been sold. At 7 o'clock on the

Tuesday morning a long queue snaked around the corner of the Arsenal stadium in north London as fans waited for their chance to buy a seat ticket at 5s 6d and 12s. It was a lot of money, but what a game was in store!

They were the two strongest teams in the country, no doubt about it. Preston had won the Cup in 1938, and since then the side had been bolstered by the arrival of several teenage sensations: Alex Fairbrother in goal and Andy McLaren and Tom Finney up front. They had yet to be beaten in 1941, having won fourteen, drawn three and scored sixty goals along the way. Top scorer was Jimmy Dougal with thirty-one, and he wanted nothing more than to celebrate his marriage in four days' time with a Cup Final goal.

But Preston were up against an awkward Arsenal side who rationed goals the way butchers did sausages. They had conceded none in the two leg semi-final against Leicester, and in the quarter-final London rivals Spurs had gone down 2–1.

At the heart of the Arsenal defence was 33-year-old Eddie Hapgood, a non-smoking teetotaller who had captained England thirty-four times. The club had paid £750 for him in 1927 and since then he had been part of the great Arsenal side of the 1930s that won five league titles and two FA Cups.

Most of Arsenal's goals came from either Cliff Bastin or the sublimely gifted Denis Compton. Compton, like his elder brother Leslie, was an all-round sportsman who had already played cricket for England and had a golf handicap of ten, a year after taking up the sport. Arsenal had spotted Compton junior at the age of 14 and had been nurturing him ever since.

Hapgood and Denis Compton were two of the reasons young Tom Finney's stomach was pulled tight as the 3.30 p.m. kick-off approached. Two days ago he'd been working as an apprentice for a Preston plumber; now 60,000 people had paid good money to watch him in the Cup Final. 'It was an absolute dream for me to be playing at Wembley,' he recalls, 'but I was pretty nervous in the dressing room beforehand.' He didn't even mind that there 'would be no medals presented to the winning side as there wasn't any gold to spare, [instead] each player from the winning side would be presented with four 15 shilling wartime saving certificates'.

A few minutes before 3.30 p.m. the teams trotted down the Wembley tunnel and out into the bright sunshine. 'As soon as I got out on to the field my nerves went,' remembers Finney. The two sides lined up facing the Royal Box, Arsenal wearing red and white shirts and white shorts and Preston with their white shirts and black shorts. The crowd quietened to a respectful murmur as both teams were presented to the First Lord of the Admiralty. A word here, a smile there, handshakes all round, and then the presentation was over. Sixty thousand voices raised the Wembley roof (which had been painted black in an optimistic camouflage attempt) during the singing of the national anthem. At 3.30 p.m. the match began.

Arsenal were into their stride from the whistle with Denis Compton using his pace to unsettle the Preston defence. Finney at outside-right had seen little of the ball when after five minutes Leslie Jones, the Arsenal inside forward, threaded a ball through to Alf Kirchen. He lashed a shot towards the Preston goal, only to see it parried by the hand of Preston's right-back Frank Gallimore. Referee Milner pointed to the spot. Penalty to Arsenal. Leslie Compton rifled the ball low with his left foot to the right of Fairbrother in the Preston goal. But the 19-year-old goalkeeper guessed correctly and his clenched fist blocked the shot. 'Compton hit the ball with such power,' recalls Finney, 'that the rebound ended up practically on the halfway line.'

The scare shook the lethargy from the legs of the Preston players. They began playing with pace and panache and precision. Jimmy

Dougal forced a sharp save from Marks in the Arsenal goal; Bob Beattie struck a shot that rattled the post and in the tenth minute Andy McLaren drilled the ball into the Arsenal net. 1–0 to Preston. 'The rest of the half was very fast-flowing and open,' says Finney who, along with the younger Preston team, hoped to wear down the ageing Arsenal side. 'We were smaller physically than Arsenal. Players like Bernard Joy and Eddie Hapgood were big men.'

Preston kept surging forward, seeking a second goal, but Arsenal stood firm. George Allison, their manager, sat in the stand and wondered for how long. His team looked jaded, as well they might. All of them had joined up and most had travelled long distances to play in the game. 'Training?' Allison had scoffed during a press conference forty-eight hours before the final. 'There is only what my men get in the services. And I shan't see many of them until they turn up at the stadium.' As a precaution Arsenal had put four former players on standby, but they weren't needed. The starting eleven made it to Wembley by the kick-off.

But Preston were punishing Arsenal's lack of preparation. Only the vast experience of men such as Hapgood and Bastin were keeping them in the game. Five minutes to half-time, thought Allison, if we can just hold out till then, we can regroup during the interval. Suddenly Arsenal launched a counter-attack. Laurie Scott, the right back, sprinted out of defence, up towards the halfway line, over it. He saw Denis Compton shouting for the ball and guided the ball to his feet. Compton rode the lunging tackle of a Preston defender, glanced up for a split second, then unleashed a thunderous rising shot that left Fairbrother in the Preston goal clutching thin air as the ball skimmed the underside of the crossbar on its way into the net. Arsenal had equalized on the stroke of half-time.

The second half degenerated into a dull tussle between two tired teams. The hordes of soldiers and sailors and airmen in the stands became restive, thinking ahead to that night's entertainment. There were a few shafts of brilliance: a Leslie Compton header that sailed just wide, a snap shot from Bob Beattie that produced a fine save and the best of the lot, a chip from Tom Finney. 'That was the most exciting moment of the match for me,' says Finney. 'It looked like it was going in but Marks managed to tip it over the crossbar.'

The referee blew the final whistle a few minutes after 5 p.m. Neither side would be walking up the thirty-nine steps to receive their savings certificates. Instead there would be a replay at a later date. For Finney the ninety minutes had flashed by in a moment. He was still in a daze when Eddie Hapgood 'shook my hand and said "well played", which was a great honour for me'.

THERE was no hurry to leave Wembley. The two extra hours allowed them by double summer time slowed down the step of the spectators as they headed out of the stadium. The blackout tonight began at 10.21 p.m. Plenty of time to find a pub and have a chat about the game over a pint of mild.

Elsewhere across London people were capitalizing on the extra daylight, especially as buses and trains were now running their summer timetables. The London County Corporation open air swimming pools – 'mixed bathing allowed throughout the day' – were open till 9 o'clock, though few people had taken the plunge on such a bracing day. South of the capital, in and around Croydon, ramblers were still striding through the countryside, a copy of the recently published pamphlet by Kenneth Ryde tucked into their back pocket. The pamphlet described a series of walks, anything from a half-day stroll to the Bird Sanctuary in Selsdon Woods, to a 16-mile route march across Epson Downs to Headley Church.

Those who preferred to take their walks in the parks of London, among the anti-aircraft guns and barrage balloons, admired the beds of pansies in Kensington Park or the trees of Hyde Park, fat with new green leaves.

On board the buses that skirted the edge of Hyde Park passengers looked agog at the interior of the alien buses, replacements from other British cities to augment London's dwindling number. The green gauze stretched tight across the windows to prevent flying glass deprived them even of the simple pleasure of watching the world go by; so people gossiped, to friends or strangers, it didn't matter. The melting of the Londoners' chilly reserve was a phenomenon of the Blitz, the unifying effect of a shared danger. And people found that they rather enjoyed being companionable to one another.

Those buses heading east up Oxford Street, waiting at traffic lights that were masked to reveal only a chink of colour, could go no further than Oxford Circus. The eastern half of Oxford Street that stretched into St Giles's Circus Road was still being repaired after the air raid of 16 April, one of eight priority roads undergoing repair. But at least the workmen still hard at work on Saturday afternoon were on schedule: two lanes of traffic would be open on Monday morning.

While the workmen filled and tarmaced Oxford Street, just behind them in the London Palladium Vera Lynn was preparing to take to the stage for the second time that day. Max Miller had finished his routine, having tickled the audience wearing a new suit made entirely of crêpe de Chine.

As Florence Desmond curtsied and took her leave, Vera Lynn waited nervously beneath the stage wearing 'a new frock which had used up all my clothing coupons'. Normally Vera liked to skip on to the stage, with girlish abandon, but at the Palladium the producer 'had an idea to have me raised up on to the stage inside a giant model of a radiogram'. She burst out of the radiogram and launched into a selection of her most popular songs: 'Yours', 'When the Lights Go on Again', 'The Lights of London' and the smash hit, 'We'll Meet Again'.

'I'd been singing "We'll Meet Again" just prior to the war,' says Vera. 'I knew it was going to do well because you get a feeling when a song is good or not, particularly if you know your audience and you know the songs they like. It was my type of song and it was easy to remember with good lyrics.'

Less than half a mile north of the Palladium, in the Queen's Hall in Langham Place, another audience was savouring the beautiful music of another era.

The Queen's Hall had been London's premier concert venue since its construction in the early 1890s. Its architect, Thomas Knightley, had built the Hall using innovative techniques: 'The walls of the auditorium will be lined with wood, fixed clear of the walls on thick battens,' he had said of his creation. 'Coarse canvas will be strained over the wooden lining, on which will be spread a film of composition, and on this will be raised ornament. The canvas is to check the vibration of the woody fibres; and as vertical forms of support internally have been dispensed with, that the sound waves may not be broken, the hollow lining, it is calculated, may be as the body of a violin – resonant.'

Since 1895 the Queen's Hall had staged the annual London Promenade Concerts, every one conducted by the country's pre-eminent conductor, Sir Henry Wood. He had never had any trouble attracting world-class musicians to the Hall; the acoustics were considered among the best in the world, Richard Strauss and Ignacy Paderewski having tested them in the 1930s. And Edward Elgar and Claude Debussy had both conducted their own compositions in the Queen's Hall.

Late on Saturday afternoon the Royal Choral Society was approaching the climax of Elgar's 'Dream of Gerontius' with the

London Philharmonic Orchestra. Malcolm Sargent, the conductor, recalled later that the music 'was what is usually known as "The Angel's Farewell". The men's voices, as if speaking for a groaning and war-stricken humanity, murmured "Bring us not Lord, very low. Come back, O Lord, how long". Whilst above them the voice of the Angel was singing, "Farewell, but not for ever! Be brave and patient, swiftly shall pass thy night of trial here, and I will come and wake thee on the morrow".'

At 6.15 p.m., when the audience at the Queen's Hall was on its way home, most of the Civil Defence personnel in London were beginning their shifts. Lew White was back at work in his sub-station in Lincoln's Inn, having listened to the second half coverage of the Cup Final on Forces Radio, and at her sub-station in the Paragon School, at the junction of the New Kent Road and the Bricklayer's Arms, Emily Macfarlane had begun her 48-hour shift. The firemen were sitting around, some making wooden toys for orphaned children, others discussing the female form. 'They were the naughtiest lot I've ever come across,' says Emily. 'One of them had three women on the go and a different one visited each day.' She took up her position in the control room, telephone by her side, waiting for the first call of the evening.

Florrie Jenkins had clocked on in the Burdett Road station and was in the control room under the watchful eye of Mr Nash, the Mobilising Officer. She liked Nash, principally because he turned a blind eye to the shoes she wore. 'I couldn't wear the proper uniform shoes because they were too big for me, so I bought a pair of boys' lace-up shoes.'

Reenie Carter and Margot Seymour-Price were on duty in the control room of 'A' Division in Westminster, both refreshed after tea and cakes in the Lyons Tea Room in Bridge Street. Across at No. 20 Dean's Yard, Canon Alan Don had returned home after lunch with Mr Wilmshurst, during which he had proposed Mayor Eaton-Smith as the People's Warden of St Margaret's. With no objection from Mr Wilmshurst, the Canon had acted quickly and arranged a meeting with the Serjeant-at-Arms on 29 May to second the nomination.

At Waterloo Fire Station Fred Cockett played darts with Pinky Petit and Tiddly May. They broke off the game to brew some Ki,* fire station slang for a mug of thick dark cocoa. The sun had dipped behind the Thames, 400yds to the north of the fire station, and dusk promised frost.

At his desk on the fifth floor of the Reuters Building, John Hughes punched the air in delight. He'd just sold a story to the *Sunday Express* about a Leeds' barber who collected his customers' clippings and sent them to the Government in the hope they might use the hair to make carpets. Hughes wasn't sure what astonished him most: the barber's idea or the *Sunday Express* agreeing to pay him fifteen bob for the piece.

Also working was 16-year-old Joe Richardson who, despite the lengthening shadows on the pavement, was still running messages for the military hatter in Basinghall Street behind Moorgate. But it wouldn't be long before he could cross the river and go out with his pals in search of some evening entertainment.

John Fowler and his 15-year-old cousin, Rose, were waiting at Camberwell Green for a bus to take them to the Peckham Odeon. *Charlie Chan at the Wax Museum* began at 7.10 p.m., followed an hour later by the main programme, *Seven Sinners,* starring Marlene Dietrich and John Wayne.

In Bermondsey, Gladys Jenner was chatting to Jimmy Gill, the best friend of her fiancé and now a Leading Air Fitter in the Royal Naval Air Service. Back home on a weekend's leave, Jimmy fancied

* A naval term, many of which were commonly in use in the London Fire Brigade as there was a long-standing tradition of former sailors joining once they'd left the navy.

a night at the cinema. The Trocadero at the Elephant & Castle was showing *Seven Sinners*. Gladys gave the same answer as all the other girls he'd asked: 'Sorry, Jimmy, we're going down the shelter tonight.'

Tom Winter and his pal Ken Humphreys were still larking about when Tom was told by Ken's mum to return home. It was 7 o'clock, the time the family gathered up their blankets and headed downstairs to Granny Humphreys' flat. They crossed Elgar Street together and as they did so the old woman stood waiting for them in the doorway of No. 53. 'Goodnight Ken,' said Tom. 'See you tomorrow.' He ran the 100yds to his own flat. In dribs and drabs people from the neighbouring flats were starting to make their way down the steps of the communal shelter, under the concrete football pitch in the middle of the estate.

At the Dockhead Fire Station Bobbie Tanner had finished the maintenance checks on the petrol lorry. Now, at 8 o'clock, she lay on her bunk in the women's quarters, a tin hut in the station yard surrounded by bricks and sandbags. The Ascot gas heater in the corner did its best to breathe warmth into the hut but the cold snuck in through the cracks. Bobbie wriggled under the blankets and continued reading. She tried to read a book a day but it wasn't always possible, particularly when some of her time was still taken up with replying to fan mail. Another letter had arrived earlier in the week, from a Mr Griffin. He'd taken the trouble to compose a few verses in her honour, though he warned her 'I'm only an amateur at writing verses so I hope you will like it'.

'To London fair city there came one day
a girl who had put her civvies away
Then came the blitz from the Nazis above
Now comes the test thought she

Or the job that I love
Though incendiaries were falling
She did not falter
She drove her lorry so the men could get water
Loaded with petrol she drove that old motor
Did she see danger, not one iota
When she arrived there did the men cheer
Now to put out that fire and then for some beer
Thank heaven we have someone to carry our banner
Three Cheers for this girl, we salute you Miss Tanner.'

At 8.30 p.m. 50-year-old Percy Goldsack was on his way home to his wife Annie in Dover, on a route he had taken many times during his seventeen years as a train driver with the Southern Railway. He stood in the engine watching the Kent countryside flash by, as fireman Stickells shovelled coal into the furnace. The 8 p.m. from Ramsgate to Dover was making good time when, with just 12 more miles to their destination, it was pounced upon by six German fighters near the village of Worth. They attacked the passenger train head-on, coming in low from the west with the dipping sun behind them. Cannon shells and machine-gun bullets ripped through the engine cab, reverberating in all directions with a terrifying thunder. Goldsack braked and the train juddered to a violent hissing halt. By the time the two men helped each other down from the footplate to the track the aircraft had climbed up into the sky and were headed back across the English Channel.

Ricochets had lacerated Stickells's body and he had wounds to his thigh, lower leg and foot. A shrapnel sliver had also sliced open the artery in his upper arm, pumping an arc of blood on to the rails. The guard, a man called Sabine, sprinted up the track from further down the train and applied a tourniquet to Stickells's arm. Then he turned to the prostrate driver. Goldsack had been hit in the chest by a bullet and death was near. There was little Sabine could do to help. A detachment of Royal Marines on a coastal training exercise arrived and one, seeing the wounded railwaymen, set off to their camp to fetch the medical officer. When he returned with the doctor half an hour later Percy Goldsack was dead, a victim of a hit-and-run raid designed to dupe the British into believing that nothing unusual was afoot this Saturday evening. Dozens of similar marauding attacks had been carried out by the Luftwaffe in the past eight months; this was just another one, and there was reason to be thankful that casualties had been so light. The British didn't know that the six aircraft were merely harbingers and that across the Channel in France, its coastline almost visible from Kent on such a clear calm evening, the Apocalypse was on its way.

9

IN the Briefing Room at Vendeville airfield Group Commander Oberst Steinweg stood before a large map of London tacked to the wall. Sitting on the hard benches, staring intently up at him were the crews of No. 5 wing of KGr53. Albert Hufenreuter sat next to his pilot, Richard Furthmann, an artless blond-haired German. Steinweg moved his pointer across the map until it came to the Isle of Dogs. 'Use this as your landmark,' he said, before listing No. 5 Wing's targets for the evening: all river bridges running west from Tower Bridge. So, London Bridge, Southwark Bridge, Blackfriars Bridge. Waterloo Bridge – easy to identify because the new bridge is being built alongside the existing one – Hungerford Bridge, Westminster Bridge, Lambeth Bridge. There were other targets, too. The factories dotted along the southern side of the Thames, the warehouses of Stepney, and the railway line that ran north from the road junction at Elephant & Castle, across Blackfriars Bridge, and into Holborn Viaduct Station.

Steinweg paused momentarily, watching his crews' faces. The X-Gerat radio beams would be intersected over West Ham, he told them, but they would be superfluous tonight. He turned and nodded to the meteorological officer, who stepped forward and began to speak. The forecast for the evening was for a bright moonlit night with no cloud cover. Such nights had once been called Bombers' Moon, but in recent months the Luftwaffe crews had come to dread the sight of the full moon. For Feldmarschall Sperrle, luxuriating in his HQ in the Palais du Luxembourg, the moonlight was still a loyal ally, and to hell with the RAF. To his bomber crews, Sperrle was known as 'The Killer'.

The rest of the briefing was like any other. Details of the light signals needed by the AA batteries on the French coast to identify aircraft, codes for the wireless operators, radio frequencies, routes to be flown, altitude height, take-off times. And remember, Steinweg reiterated to the bomber crews, once you are over London you have fifteen minutes to find and bomb your target.

Benches scraped to the straightening legs of the airmen. Steinweg wished them luck as they filed out of the briefing hut and made their way to draw their parachutes and collect their flight rations. There was little chatter as the men grabbed packets of dry raisins and bars of chocolate laced with caffeine and stuffed them into the pockets of their flying suits.

Hufenreuter retired to his room to plot his course. Of his crew of five, he was in charge, senior in rank even to Furthmann, his pilot. The official route, the one laid down by Steinweg, would take No. 5 Wing north from Lille over Belgium and the western edge of Holland before they swung sharp west towards England. To Hufenreuter, this route was stupid and dangerous. They would be flying over North Foreland and following the Thames from its estuary right into the heart of London. It didn't take a combat veteran of twenty missions to see that on a clear moonlit night such a route was asking for trouble. Hufenreuter poured himself a coffee and plotted a new course.

The eleven crews of KGr100 selected to fly sat down to supper in the dining room of the Hotel le Bretagne in Vannes. A quiet meal together was part of the pre-mission ritual, an opportunity to derive strength from one another. Like any group of young men thrown together under pressure, they each coped with the stress differently. There was the crew's quiet man, who ate his meal in silence, smiling at the wisecracks of the crew joker; there was the serious man, intense and earnest, who lectured the youngest member of the crew on life and love. The youngest member of the crew sat there, wide-eyed with excitement, not really listening, pushing his food nervously round his plate.

Shortly before 9 p.m. the crews started to drift out of the dining room. They had about an hour to kill before the short drive to the airfield. Some men lay on their beds, hands behind their heads, staring listlessly at the ceiling; others wrote letters home, read a book, flicked through a magazine. In the corner of the dining room, a game of skat was under way, with three players waiting patiently as the dealer flicked them each ten cards. Whatever the airmen did, there was one thing they had in common. From time to time, with increasing frequency, they glanced down at their wristwatches.

IN the West End of London, cinemas were turning out customers at 10 o'clock. The day before, the Metropolitan Police had agreed to the new closing time, an extension that was a nod to the double daylight saving. Now Londoners had twenty minutes to get home before the blackout. At the Trocadero in the Elephant & Castle, the curtain had just gone down on the Twelve Tiny Tappas and the Sid Kaplan band. At New Cross dog track the last race had finished an hour ago, but punters were still slaking their thirst in the pubs outside. At a pub in Pentonville an argument had spilled out on to the street. Constables Soames and Whitehouse arrived to break up the trouble. But Elizabeth Rogers, clearly the worse for wear, showered the two policemen with abuse. When her daughter, Eva, threw in some choice words of her own, the pair were arrested. At the Havana Club in Denman Street the band had yet to take to the stage but the private guests were already clinking champagne glasses. All over London it was an ordinary Saturday night.

WHILE the Luftwaffe aircrews lived and breathed by the ticking of their watches, their ground crews worked furiously to ensure all 370 aircraft of the first sortie were ready. Busiest were the bomb loaders, sweating profusely in the chill night air. German high explosive (HE) bombs in 1941 ranged in weight from 50kg to the 1,800kg beast that was known to the boys of Bomb Disposal as 'Satan'.

Luftwaffe HE bombs came in two casings, the thinner-cased Sprengbombe-Cylindrisch (SC) or the thicker-cased Sprengbombe-Dickwandig (SD). The SCs were 55 per cent explosives and the staple bomb of the Luftwaffe during the Blitz. Comprising three thin-cased sections, they had a solid nose and an alloy tail. Many SCs had a ring (*Kopfring)* fastened to the nose to stop them boring too deep into the ground. They were intended to cause maximum blast effect and, more often than not, they were successful. 'Satan' had a thin sheet steel casing and when one was dropped in London for the first time, in Great Dover Street in April 1941, a terrified eye-witness described it as a 'gigantic sheet of flame'. A man walking half a mile away was knocked insensible by the blast.

The SD bombs, 35 per cent explosive, were semi-armour piercing fragmentation bombs of malleable cast steel, capable of catapulting shards of shrapnel over a radius of 1,000 metres. The deadliest SD bomb was the 'Fritz', weighing in at 1,400kg and standing a couple of centimetres under 2 metres. Such scourges were used sparingly. Most of the German bombers carried bombs of 1,000kg or less because the lighter the bomb the more they were able to load on to the aircraft. The older Heinkel carried its load internally in eight vertical bomb chutes which could each accommodate one 250kg bomb or four 50kg bombs. They were all attached nose upwards, secured in position by means of an eye bolt mounted on the casing. The bomb loaders who worked feverishly on 10 May took no notice of the variety of devices strapped to the four tail fins of some bombs.

Some of the fins had not been tampered with, but on others small cardboard 'organ pipes' had been riveted on. When the bomb left the aircraft on the start of its 500mph descent, the air rushed through the pipes to produce a noise that curdled the blood. 'Screamers' were what the British called them; a scream that burrowed its way deep into the sub-consciousness of thousands. Their shrieks also allowed delayed action bombs to drop unheard, and to explode any time up to eighty hours later.

Other fins had incendiary bombs strapped to them, sometimes one, sometimes two, all with the intention of creating maximum chaos on the ground. Occasionally, one of the bigger bombs, a 500kg or a 1,000kg 'Hermann', wore six magnesium panels round its body like a necklace. Five incendiary bombs were clipped on to each of the panels. They looked peculiar when they were being loaded into a Heinkel in the still chilly French evening, but as soon as they were dropped from 12,000ft, the whole necklace was released from the bomb and then, one by one, the six panels detached and plummeted earthwards with their five incendiary bombs clipped on.

Incendiary bombs were used in greater number than the bigger but less effective oil bombs. These were over 1.5 metres in length and weighed 113kgs. The casing held approximately 65 litres of oil, sometimes a mix of petrol and oil. But the disadvantage of oil as an incendiary was although it emitted twice as much heat as magnesium it was lighter and the bombs rarely penetrated buildings. Mostly they exploded on the roof, releasing a film of oil over an area 20 metres in diameter and as oil burns upwards they caused few serious problems for the firefighters, who used foam to douse the flames.

BY the time the eleven crews of KGr100 arrived on Vannes airfield their aircraft were fully loaded. As Beleuchter-gruppe (fire-lighters), they carried almost exclusively incendiary bombs. One or two of their

Heinkels had a 250kg bomb in their chute but in most were what had come to be known as incendiary 'breadbaskets'.

The moniker 'breadbasket', like 'screamer', was a product of the British sense of humour; their belief that by ridiculing something nasty it diminished its potency. And the 'breadbasket' like the 'screamer' was very nasty. The 'breadbasket' was actually a three-panelled aluminium container measuring 43ins and painted, like the thirty-six incendiary bombs it carried, the standard German grey green. A central rod held the three panels together and the container was fitted with a clockwork mechanism that released the rod five seconds after leaving the aircraft.

A solitary incendiary bomb was a straightforward piece of ordnance. In size and shape it resembled the truncheon of a London policeman, and its 1kg weight was also similar. A British Government handbook at the start of the Blitz described an incendiary bomb as 'designed to give an intensive combustion at a high temperature by burning the body of an electron which is ignited by the thermite'. To the average slum dweller in Stepney or Southwark, it was all gobbledygook. All they knew was that for a few moments an incendiary bomb burned very brightly and with a strange 'psss-psss-psss' sound.

Initially the British hadn't shrunk from the threat of incendiary bombs. During air raids, teenage boys, young women and middle-aged men felt that by chasing and extinguishing incendiaries they were 'doing their bit' for the war effort. Across the capital, Londoners ran up and down streets dousing fizzing incendiaries in water or extinguishing them with sand. Frequently Air Raid Wardens had to break up scuffles between people arguing that it was 'their' incendiary bomb, they had found it first, and they had the right to put it out. Word soon reached the Luftwaffe of this British bravado and towards the end of 1940 they began screwing small explosive charges, about 1¼ins long, into the tail of some incendiaries. The charge detonated three to five minutes after the incendiary had started to burn. It rarely killed the plucky civilians, merely ripped off an arm or a foot. But practically overnight people became less willing to chase falling incendiaries.

The 'breadbasket', like the small explosive charge, was a further evolution in the development of the incendiary bomb. And as the crews of KGr100 prepared for take off shortly after 10 p.m. most of their aircraft carried in each of their eight bomb chutes four 'breadbaskets'. Four breadbaskets in one chute equalled 144 single incendiary bombs. Multiplied by eight, and each Heinkel that taxied down the grassy L-shaped runway at Vannes was carrying 1,152 incendiary bombs. The canisters had all been pre-selected, which meant once they were over their target the navigator (also the bomb aimer in a Heinkel) wouldn't have to concern himself with selecting the time interval for release. Once he pressed the bomb release button one canister would fall every 50 metres. As the last of the thirty-two dropped, the crew would be able to look down and know that they had just started a 1,600-metre ribbon of fire.

10

ALBERT Hufenreuter arrived on Lille North airfield at 10.10 p.m., the time KGr100 were climbing into the air to launch the raid. Their cigar-shaped Heinkel was silhouetted in the bright moonlight, its sleek lines unimpaired by 600lbs of protective armour. Most of the plane had been daubed in black paint to make it less visible to British night-fighters. Even the markings – the swastika on the tail fin and the German cross on the side of the fuselage – had been camouflaged, and the 1,000kg bomb that was clamped to the Heinkel's undercarriage, like some giant bloodsucking leech, was also coated in black. Four incendiary breadbaskets had also been loaded in the internal bomb chutes.

The newer versions of the Heinkel, as well as carrying their bombs externally on a PVC rack, had twin Jumo engines instead of the Daimler Benz ones in the older models. These were both being run up as Hufenreuter and Richard Furthmann, the pilot, climbed on to the wing and slid down into the cockpit through the sliding entry panel. The other three crew members ducked under the fuselage and climbed up into the belly of the Heinkel through the gondola. It wasn't easy, squeezing up through the gondola on to the narrow passageway. Their fur-lined flying suits, ribbed-pattern life preservers and parachutes restricted movement. As the crew waited one by one to squirm their way up through the gondola, all around them was the deafening roar of engines being revved. Then the first plane began to thunder down the runway, the rest to follow at three-minute intervals.

Josef Berzbach, the unflappable mechanic, wriggled up and moved awkwardly along the passageway to the cockpit. During the take-off Berzbach squatted down between the pilot and the navigator; once airborne he took on his secondary role of gunner in the lower fuselage gondola. The gondola was nothing more than a hollow shell, utterly exposed to fire from a British fighter. Heinkel crews called it the *Sterbebett* (death bed).

Last on board were Eggert Weber, the dorsal gunner, and Karol Gerhardt, the callow wireless operator. Weber climbed into his turret and strapped himself into his swivel seat, peering out of the glass dome that housed his 7.9mm MG15 machine gun. Once they were airborne he felt only marginally safer than Berzbach, lying on his death bed. The dome was enclosed during take-off but in the air Weber slid back its rear end to operate his machine gun. The turret's design, however, restricted his field of fire to 180° and with only a wafer-thin shield of armour in front of the gun, Berzbach was vulnerable to a determined and skilful enemy who came out of a coal black sky.

Beneath and slightly behind Weber in the radio compartment, Gerhardt clipped himself into his sling seat. His flying helmet was made of a lightweight material, not leather, with two large black earphones sewn into the material. It fitted tightly and smoothly over his skull and came down right over his forehead to finish just above his eyebrows. Gerhardt was another crew member proficient in two skills; ostensibly he was the wireless operator, but the two single MG15 machine guns on either side of the fuselage were his to man if need be.

In the cockpit Furthmann slid into his pilot's seat, the hard maroon leather sighing under his weight. He ran the intercom cable down his back and clipped it to his flying suit. On his left side was a box frame containing the engine throttle, airscrew-pitch control levers and fuel tank selector and oil cooler flap control levers. On his right side were the elevator trim wheel and rudder trimmer. The control column was centrally positioned between him and Hufenreuter's seat on his right. The U-pattern pilot's handgrip was at the end of an extension arm that could be swung to the left or, in the event of the pilot being incapacitated, to the right, in front of the navigator.

With the five airmen at their stations, Furthmann pushed the throttles wide open and the needle on the rev counter scampered round the dial. The port engine gave a deep, throaty, satisfied roar. Its starboard twin did the same moments later. Weber, in the upper turret, was engulfed by a billow of thick blue smoke. Furthmann ran through his checklist, looking up at the instrument panel suspended from the roof above his head. There was no logical scan pattern to them; the engine instruments were to the right of the large magnetic repeater compass and the flight instruments to the left. He glanced round behind him at the oil/fuel pressure gauges and the warning gauges that ran along the rear cockpit roof: flaps up, undercarriage locked, fuel good, oxygen ready. He hooked his own oxygen mask on to his flying helmet so that it rested on his crown.

He called up the crew one by one on the intercom. Everyone was ready, their faces pale with apprehension. 'OK,' said Furthmann, 'standby for take-off.' He flashed a signal to his maintenance mechanic who ran forward and whipped away the blocks from in front of the two wheels. The red lamp in front of them turned green. Furthmann flicked off the brakes and pushed the throttles gently forward.

Hufenreuter was sitting next to his pilot on the fold-up seat, looking straight ahead out of the curved glazed panes of the Heinkel nose. It was like peering down a glass tunnel. Once they neared their target he would leave his seat and crawl forward on his stomach to the lower section of the nose and squint through the bomb sight equipment. In the event of an enemy attack, Hufenreuter would also man the 7.9mm machine gun in the nose, but that was unlikely. The RAF nightfighters liked to stalk their prey from the rear.

They gathered pace down the tarmac runway of Lille North, Furthmann applying just the right amount of throttle, careful not to bring back the stick too early. As they neared the maximum of 1,400 horsepower Hufenreuter thought momentarily of the 1,000kg bomb beneath them. Ever since Heinkels had begun carrying bombs externally there had been a number of crashes during take-off caused by drag despite its normally reliable handling. He could feel the sweat on his palms inside his flying gloves. Faster. Faster. Faster. Then the charge of exhilaration as they climbed. Hufenreuter glanced at the luminous dials on his wristwatch. It was 10.30 p.m.

11

AT 10.30 p.m. Tom Finney and his Preston North End teammates were crowded round the bar of the Northampton hotel where they were staying for the night before catching the train home on Sunday. Pints of light and mild oiled their tongues and 20yd dribbles up the centre of the pitch were doubled in distance. 'There was no after match function laid on for us after the game,' recalls Finney. 'We bathed, changed and took the bus back to Northampton where we had a meal in the hotel.'

At around the time referee Milner had blown the final whistle at Wembley, a telephone had rung in Fighter Command Headquarters in London. It was No. 80 Wing, the RAF's 'Hush-Hush' boys based at Harrow-on-the-Hill. Their monitoring unit had detected two German beams intersected over West Ham. It was just a test, Fighter Command was told, and the beams had now been switched off, but there might be something in the offing tonight.

The order to test the X-Beam had come from General-lieutenant Willy Haenschke, the Luftwaffe Senior Signals Officer in northern France. KGr100 would be flying down the beam in a few hours and it was standard procedure to check that it was functioning correctly. He wasn't concerned that the British would pick it up. They tested the beam on most nights and other transmitters along the French coast would later lay decoy beams to confuse the British monitors.

The news made little difference to Air Marshal Sholto Douglas, head of Fighter Command. With a full moon tonight, the first since 11 April, and no German attack on London for three weeks, he had prepared for a 'Fighter Night': 74 Squadron had been moved from Gravesend to West Malling and if there was a raid he knew he could call on over a hundred fighters. As the day fighters had no radar on board, the Spitfires, Hurricanes and Defiants would patrol in layers at altitudes between 14,000ft and 23,500ft, and at separations of 500ft. The anti-aircraft guns dispersed throughout London and its suburbs, dotted along the banks of the Thames, in parks and on playing fields and on top of lorries that moved from one quiet residential street to the next, had orders to fuse their shells to explode at altitudes no higher than 12,000ft. While the 'Layer' operation was in progress, the night-fighters would patrol the coast and the rural areas of Sussex, Kent and Essex and up towards the Wash. This way the day fighters knew that any aircraft straying into the designated 'Layer' zone shouldn't be there.

The routine dusk patrols had been carried out before the sun set at 9.36 p.m. Nothing to indicate a large build up of bombers had been reported along the coastlines of southern and eastern England, save for the six bandits that had attacked the Dover-bound train.

But at the time Hufenreuter and his crew took off from Lille North, the British radar stations had started to pick up the blips of the KGr100 Fireraisers approaching the French coast. The information was passed to Fighter Command's Underground Filter Room, where WAAFs (Women's Auxiliary Air Force) listening on their headphones stood around a large map of the English coastline, croupiers' rakes clutched in their hands.

From his position on the circular gallery, Douglas looked down as the WAAFs used their rakes to push the magnetic iron markers –

'Hostiles' they called them – across the Channel towards the south coast. Next door to where the course of the Germans was being plotted, in the Operations Room, calls were made to Fighter Command's Sector Operations Rooms peppered all over the south of England. 'Hostile Aircraft approaching.'

Up in the gallery the Air Raid Liaison Officer passed on the news to Scotland Yard, where it was relayed to London's scores of police stations. Across the 700 square miles of Europe's biggest city, the air raid sirens started to howl. Like the sound of a woman mourning the death of her child, the ululation began slowly, then rose in pitch to a crescendo. And then, as if she was pausing for breath, it diminished for one, two, three, four seconds, before rising again to its unnerving climax.

IT was the responsibility of each individual London borough to sound the air raid siren once Scotland Yard had alerted them. Some boroughs acted more quickly than others. In Croydon the siren started its minute's screech at 10.53 p.m. In Westminster it was 11 p.m. Joan Veazey and her husband, Christopher, heard the siren sound in Kennington at two minutes past eleven. 'That awful warbling-like sound of a howling wolf,' Joan wrote in her diary. To others the air raid siren was like a 'banshee' with its 'dreadful howling notes'. For some Londoners, the same ones who talked breezily of incendiary 'breadbaskets', the air raid siren was a person, 'Wailing Willie' or 'Wailing Winnie', depending on your gender. Others just called it 'The Wobbler'.

Families with pets knew the air raid siren was coming by the reaction of their cats and dogs. A good minute or so before the sound was audible to human ears, an animal could hear the siren in the neighbouring borough. Dogs barked at the window and cats shot out of the room to some dark corner of the house where they felt safe. Their owners exhibited less fear. Some headed to the Anderson shelter in the garden or backyard, but many others waited for the wailing to stop, then went back to sleep or carried on reading or finished the crossword in the newspaper.

IN No. 136 Redriff Estate there was little room for pets. Tom Winter, his two elder brothers and two younger sisters and his mum were putting up Mary Nunn and her family from upstairs for the night. Tom's dad, Harry, was out on air raid duty, which made for one less, but there was hardly room to move in the cramped front room. 'The sirens began to sound at 11,' recalls Tom, 'and even before the wailing warning had died down you could hear the anti-aircraft guns in the distance.' Tom's mum and Mrs Nunn craned their head to the wireless to catch the last twenty minutes of the Saturday night play, *The Extraordinary Conduct of Bridget*, while the children 'amused ourselves with card games and the new game of Monopoly that had just come on the scene'.

In Barking, Vera Lynn was already back home when the warning sounded. During previous night raids, when daylight hours were fewer, the siren had interrupted her performance. 'Some would leave,' she recalls, 'but enough stayed for us to carry on with the show. Then when the show had finished we'd all have a little sing-song and some people would get up on stage and do a song of their own.'

On such nights Vera remained in the Palladium until the All Clear sounded, 'sitting on the floor of the passage way that led from the stage door to the dressing rooms with my back against the wall. I was by myself, the rest of the cast had either gone home or gone down the Underground'. But at 11 p.m. on this night Vera was tucked up in Upney Road, though how much safer she was at home than cowering in the Palladium was a moot point. Three weeks earlier the Luftwaffe had dropped fifty-four parachute mines on the borough of Barking, killing forty and levelling several streets. But people still decided to 'chance it'.

Half a mile south of Vera's home, John and Selina Curtis were having a family party at their house in Keith Road; their eldest son Alf and his wife Ethel were round for supper and with their other son John, 23, and their 19-year-old daughter Elsie, it was a full house. Normally the Curtises went to the communal surface shelter which had been built on a strip of wasteland between Keith Road and Greatfields Road. Architect of the shelter was 45-year-old Job Drain, one of Barking's most famous sons. He had been awarded a Victoria Cross in August 1914 – the first Barking recipient of the medal – and on his return the borough rewarded Drain's bravery with the presentation of a purse of gold and a watch. Having survived one war without a scratch, Drain was determined to do so again. He organized the construction of a shelter that consisted of five Anderson shelters laid side by side and covered with a couple of feet of soil. The forty people inside felt they were in a tunnel and the presence of a VC holder stiffened their sinews. When the siren went at 11 p.m., Drain and his wife, Patricia, ducked down into the shelter. The Curtises continued with their family party.

All over the city people reacted differently to the siren. In Maida Vale, West London, 29-year-old Olive Jones was in bed after a hectic day delivering rations to the crews of barrage balloons. She'd returned home to discover that her cat, 'Spitfire', had just given birth to a litter. 'The alert sounded sometime around 11 p.m.,' she wrote, 'just as I was on the verge of dropping off to sleep.' Olive curled up under the eiderdown and tried to get back to sleep.

In Shepherd Market, a quaint mews in Mayfair where eighteenth-century servants' houses overlooked cobbled streets, Madeleine Henrey and her husband, Robert, calmly wheeled the cot containing their baby son, Bobby, into 'a corner of the entrance hall between the front door and the living room' when the siren went. Madeleine then took her favourite crystal lamp with its pink silk shade and placed it on the floor between her and the cot. There they sat in the semi-darkness holding hands, like two teenagers in the cinema, while Madeleine gently rocked the cot back and forth. It was almost a year ago that they had fled their idyllic life in Normandy – she keeping chickens, he growing cider apples and both writing articles for magazines – and returned to England, birthplace of Robert. He had met Madeleine at the Savoy Hotel in 1928, stealing a glance or two over a nail file as the pretty 21-year-old Frenchwoman gave him a manicure. They had little in common, she was the daughter of a miner from northern France and he an old Etonian, but they were married within months. She quit her job at the Savoy and helped her husband with his career as a gossip columnist, employing to great effect her innate élan and *savoir-faire* as radars which picked up tittle-tattle for Robert. It was a good life but one they swapped for the rustic charm of Normandy in 1937. When the Germans marched into France three years later the Henreys fled to England.

In the American Bar of the 500-room Savoy Hotel, barman Arthur Massara cleaned glasses and bantered with the unconcerned drinkers. Massara had been a wine waiter in the restaurant before the war, going about his business in obsequious and unobtrusive fashion. Then one day in the later summer of 1940 an American journalist booked himself into the Savoy. Another followed, then another, until a couple of months' later, the hotel was full of them. There was mutual antipathy at first, a wariness born of ignorance and stereotypes, but the Blitz replaced it with admiration for the way both sides conducted themselves: with courage and resilience. American journalists coined the phrase 'London Can Take It', and cabled home reports of cheerful Cockneys and stoical aristocrats and golf clubs 'where the position of delayed action bombs would be marked by red flags at a reasonably safe distance and that a ball removed by enemy action might be replaced without penalty'.

The Savoy appointed Jean Nicol as the hotel PR manager. She became all things to the journalists: mother, nursemaid, confidant and fixer. No problem was too great for her, she even finessed the transformation of a small, inner, windowless hotel room into a residents' bar that opened at midnight and closed at 8 a.m.

But the preferred drinking hole for the journalists was the American Bar. It became so popular that Massara was transferred from the restaurant to run it. Small and quick-witted, he was an instant hit with his customers. They christened him 'Titch' and thereafter it was 'Titch's Bar', and 'Titch's Special' was a three-decker toasted sandwich of scrambled eggs and bacon.

On May 10 a handful of journalists were enjoying Titch's hospitality. Young Ben Robertson of *P.M.*, a New York tabloid, was drinking in the corner with an English colleague; the silver-haired bachelor Jamie Macdonald of the *New York Times* and his boss, acting chief Bob Post, plump and bespectacled, were playing stud poker; and of course Larry Rue of the *Chicago Tribune* was present. Rue was one of the Savoy's favourite guests. He had a wisecrack for every occasion and Nicol, though she loathed the *Tribune* with its view that America should leave Britain to fight its own battles, described Rue as 'the most lovable' of the foreign correspondents. 'Larry was no isolationist,' she said. 'He was frankly pro-British and solved his problem by sticking to the news in his coverage.'

Rue had filed his copy for the Sunday edition of the *Tribune* earlier that day, with the accompanying headline, 'An Air Blitzkrieg Breeds Crime, London Learns'. It was a typically gritty piece by Rue, who avoided sentimental cliché-ridden stories, and simply reported the facts. The opening paragraph began: 'An increase in crime in the London area since the air blitz began last September was revealed today. A report on the work of the Met Police by Sir Philip Game, police commissioner, showed: Juvenile crime on the increase, accounting for 48 per cent of all arrests ... 4,584 cases of looting. Game said: "The most distressing feature has been the number of cases in which members of the various public services abused what is in fact a position of trust."'

Nicol always knew where Rue would be if she needed to find him. 'He was the most regular habitué of Titch's Bar,' she recalled. When the air raid siren sounded this night he barely looked up from his chess board. His opponent was the actress Claire Luce, a beautiful blonde whose long legs were stretched out under the table. Those legs had danced with Fred Astaire in the 1930s and she was now in London working. She felt reassured by Rue's sang-froid. He himself had recently written that 'familiarity is breeding a certain amount of contempt as regards night bombing', and he was living proof.

The Savoy's restaurant had been moved from underneath its glass dome to the bombproof River Room, where guests sampled the ingenious dishes of French chef Latry, among them tarte de Gaulle and Woolton Pie, created in honour of the Food Minister. In the corner, on Saturday evening, house pianist Carroll Gibbons was tinkling the ivories of his white piano. Gibbons had been back in the States visiting relatives when war had broken out. He cut short his trip, much to the amazement of American officials who wanted to know why he was so keen to get to England. 'Because the people over there have been nice to me and this is no time to run out on them,' he replied. Once already in the Blitz, Gibbons had been thrown from his dais by one close-landing bomb. But there were compensations; he had accompanied Noel Coward on the piano and each night the River Room's dance floor was packed with the rich and famous.

Gibbons didn't miss a key as the siren sounded. Nor did its lament disturb the two men with lived-in faces who were planning a Sunday cycling expedition to Kent over a meal and a bottle of wine.

Quentin Reynolds and Ed Beattie were members of Jean Nicol's press pack. The 39-year-old Reynolds was one of the Savoy's most faithful customers, certainly the journalist with the biggest expense account: £200 a week wasn't unusual for Reynolds, whose largesse matched his courage. He had spent many weeks on the French front in the early summer of 1940, filing reports for American readers, before driving to Bordeaux and using his ancestral Irish blarney to wangle a passage on one of the last ships to sail for England. Reynolds was a familiar sight to the housemaids who came to clean his suite; propped up by pillows, with a typewriter between his legs, a bottle of 'old fortifier' by his side, Reynolds bantered with the maids in Cockney rhyming slang and warned them of his three pet goldfish in the bidet.

As the siren finished Reynolds looked across the table at his companion, Ed Beattie of the United Press Agency. Beattie shook his head with casual indifference. 'There's that nasty man,' he said. 'No one paid any attention to it,' said Reynolds. 'The minutes passed and we'd forgotten the siren.'

At their home in Lord North Street, Westminster, 52-year-old Walter Elliot and his wife Katharine were preparing for bed, having earlier hosted an Indian Army officer for dinner. Mrs Elliot recalled, 'there was a tremendous outburst of air raid warnings … instead of going into the air raid shelter, which we had in our own house at the time, he put on his uniform'. Her husband was the Conservative Member of Parliament for the Kelvingrove Division of Glasgow and Director of Public Relations at the War Office. His life had been shaped by his service as an infantry officer in the trenches during the First World War (in which he won an MC and bar), an experience that led him into politics. In the early 1930s he had been touted as a future Tory Prime Minister but when Neville Chamberlain succeeded Stanley Baldwin in 1937 Elliot found his star waning under the new leader.

After dinner Elliot had taken his guest on a tour of Westminster Palace. He explained that Parliament wasn't looking its best; earlier raids had shattered the windows, the sword on the equestrian statue of Richard the Lionheart had been bent and a wall of sandbags protected the east side. Inside the Palace there was a melancholic air; a string of hurricane lamps lit up the Central Hall during the blackout and every bleak gap on the wall betrayed another painting or tapestry that had been removed to safety. But the House had lost nothing of its symbolism and the Indian officer had left well satisfied.

The last wail of the siren still clung to the night air as Walter Elliot grabbed his helmet and dashed out to report for duty.*

Churchill was out of town this evening, having travelled up to Ditchley Park in Oxfordshire on Friday afternoon. He'd departed in a state of high excitement, according to John Colville, his Private Secretary. Operation Tiger, the shipment of tanks through the Mediterranean to enable the British to take the offensive in North Africa, was on course. 'The convoy has passed through the narrows between Sicily and N. Africa,' Colville wrote in his diary, 'and only one ship out of five, carrying a huge consignment of tanks for Wavell [General Sir Archibald, Commander-in-Chief in the Middle East], has gone down ... the P.M. thinks this will have far-reaching consequences on the war.'

Ditchley Park was a sumptuous Georgian Mansion, built in 1722, and owned since 1933 by Ronald Tree, the American-born Conservative MP for the Harborough Division of Leicestershire and former editor of *Forum* magazine, and his beautiful Virginian wife, Nancy, a niece of Nancy Astor. The Trees had put Ditchley at the disposal of Churchill six months earlier, after bombs had dropped disturbingly close to Chequers, the traditional country retreat for prime ministers. Was it a fluke or had the Luftwaffe identified the house and its long gravel drive in the bright moonlight? 'The P.M. is

* Firewatching shifts began at 8 p.m. and lasted till 6 a.m., with each pair of watchers doing a two-hour shift before being relieved by the next pair. During a heavy raid, however, most firewatchers mucked in, regardless of their shift hours.

obviously worried about the possibility of an attack on Chequers,' wrote Colville, 'and says that he does not object to chance "but feels it is a mistake to be the victim of design". With the evolution of the German "beam" methods, he fears that accurate blind bombing may become a reality.'

Thereafter Churchill repaired to Ditchley Park when the moon was high. His youngest daughter Mary recalled that after the fortifications of Whitehall and Westminster they were charmed by Ditchley and 'gazed with keener appreciation on elegance and beauty, and glowing, lighted interiors'. Guests found the Italian garden an idyllic refuge in which to forget about the Blitz.

As the 'Wailing Winnie' shattered the peace of London's evening, Churchill and some of his closest advisors were discussing the latest setbacks in the Balkans and the Middle East. General Sir Hastings Ismay, Professor Fred Lindemann and Brendan Bracken puffed on cigars and fingered the stems of their brandy glasses. Churchill had a glass of champagne, while his bodyguard, Detective-Inspector Walter Thompson, was in the baronial hall making sure the film show was ready for the Prime Minister. Tonight he'd be watching *The Marx Brothers Go West*.

If there was fear anywhere in London as the sirens sounded, it was south of the river, in Peckham, Camberwell, the Elephant & Castle and Bermondsey. There was fear among the working class, who had suffered so much already and who had no second home to go to. These were the ones who could do nothing but 'take it' on London's behalf.

The demure Gladys Jenner and her 5-year-old brother, Brian, wished their dad was with them when the air raid siren went over Bermondsey, but he was in Buckinghamshire visiting Gladys's younger sisters. Gladys, too, would have liked to have been evacuated from London. 'I was a coward,' she says. 'I didn't want to be a hero and I didn't want to stay in London. I would've given anything to have gone.'

The dilemma for Gladys and her mum was whether to go to the public shelter under the railway arch by the Peek Freans factory, 'where people made beds out of empty orange boxes', or stay in their own house. It was a cold, cold night and Gladys shivered at the memory of those winter nights under the arches. They decided to sit tight and see if the raid was a big one or not.

John Fowler and his cousin Rose were walking west along Peckham Road, grizzling about their evening's entertainment. They'd seen *Charlie Chan at the Wax Museum*, and the *South London Press* reviewer was right, it was 'very peppy'. But halfway through the main feature, *Seven Sinners*, the picture had flickered and died. As people swivelled in their seats and glared towards the projectionist, the manager walked to the front of the theatre. 'I think it's going to be a very bad night so we are going to close the cinema down'. There was a gurgle of discontent as friends and lovers looked at one another in irritation. 'Don't worry,' said the manager, raising his hands as if he was surrendering. 'We'll give you tickets for another show or another night, but I think it's best if we shut down.' John and Rose moved reluctantly from their seats towards the exit. 'There'd been no air raid siren,' remembers John, 'but the manager sounded scared.'

John and Rose milled around the outside of the cinema for a few minutes, chatting to one or two old friends and telling jokes about the peculiar habits of the country folk. Then they started walking along the Peckham Road towards Rose's house on Vestry Road. Wasn't it extraordinary, they remarked, the brightness of the moon? Normally they would be groping their way painfully along the road, relying for direction on the white bands painted around trees and the white kerbstones. But tonight they could see the road, the pavement, even the dark sinister silhouette of the Camberwell Mental Hospital that stood on either side of Peckham Road. It was just before they turned left into Vestry Road that the siren went. 'When we arrived at Rose's house, her dad was waiting for us,' recalls John. Rose's dad, Stan, had a bakery in Kirkwood Road, the other side of the Peckham Odeon. John was struck by the worry in his eyes. Normally Stan was a happy-go-lucky bloke, but now he 'had a feeling something bad was going to happen'. He ushered his daughter inside, then turned to John. 'I think you'd better go home, son, it's going to be big tonight.' John said goodnight and began walking back towards his parents' house in East Surrey Grove along the Peckham Road.

In St Mary's Road, less than a mile east of Peckham Road, Gladys Shaw was about to leave on some missionary work. Now with the alarm wailing she found herself battling a feeling of dread. 'It was having to go out in a raid on one's own that was bad,' she says. 'I always wished I had someone with me.'

The parishioners of St Mary's had sheltered in the crypt of the church when the Blitz started, preferring the security of the Lord to the street shelter. But in common with scores of other London churches, St Mary's was soon hit. On 22 September 1940, a large bomb had shattered the church, demolished the vicarage and taken out a dozen of the neighbouring houses. The house where Gladys rented rooms was also hit, 'but they let me stay on in a basement room of this blasted house'. Workmen came in and boarded up the windows and repaired what they could of the walls with thick paper and battens, then they wished Gladys luck and departed.

There were, says Gladys, 'about 20 street shelters in our parish and in an average night I would visit seven or eight. I stayed for about 20 minutes in each one, asking how everyone was and talking to the children if they were scared. But the people really looked after themselves'.

After she'd cycled back from her picnic at Keston fishponds Gladys had searched her Bible selecting a suitable passage to take into each shelter. 'I read the same Bible passage in each shelter but changed them every night. Mostly they were words of comfort, reassuring the people that God was with us whatever happened.'

The fellowship in the shelters rarely wavered. If it did, it was a hostile question thrown in the direction of Gladys. How could she talk about peace when those German bastards were dropping bombs? And how could God stand by and let the innocent suffer? 'Why did God allow all the pain was the question I was asked most,' says Gladys. 'And my answer was always the same: it wasn't God waging war, it was man.'

But it was futile to argue with people unhinged by fear and anger. Gladys had lived alongside suffering for years; during a meningitis outbreak in 1938 she had watched mothers bury their babies. Now she was witness to a new suffering, more sudden and violent, perhaps, but working in the first aid post she had seen women brought in with their legs off. 'I never thought to question my faith,' she says, 'at any time. I just thought that the war was wrong but that we were justified in defending ourselves. I accepted the suffering and tried to preach a message of peace; that it was better to love your enemy than hate him. I remember one old woman whose roof was blown off by a bomb as she lay in bed. "I've forgiven that Hitler," she told me. "Yes, I forgive him. But may God give him his due".'

As Gladys walked out on the street the ugly symphony of AA guns on Peckham Rye Common began firing. At the same time 16-year-old Joe Richardson was on a tram going down the Walworth Road after an evening with his pals at the Trocadero at the Elephant & Castle. 'About halfway down,' recalls Joe, 'the driver said "Right, that's it, I'm stopping here", and he left the tram in the middle of the road.' Joe jumped off opposite East Lane, where the Saturday market had ended a few hours earlier. He started walking down the Walworth Road, then stopped. The noise was unmistakable, though to someone new to London it was just a distant uneven drone that might have come from some far-off factory. People began to run in panic, like a small creature that knows itself to be the prey of a larger one. They ran towards the Elephant & Castle or to Camberwell Green. Some sprinted to the brick surface shelters, while others huddled in the doorways of shops, tobacconists, greengrocers, tailors, their limbs leaden with fear. 'There was a sense of panic that hadn't been there on other raids,' recalls Joe. 'People had just had enough of it all and they started running. I didn't believe in running because you're just as likely to run into trouble as you are to run away from it. And anyway I was 16 and at that age you think everything will be fine.'

12

AT the controls of the eleven Heinkels of KGr100, the Fireraisers, now 15,000ft over London, the pilots continued to run one engine slightly faster than the other. This was the uneven drone Londoners recognized as uniquely German. The noise of the RAF was sweet and true, but Luftwaffe bombers wheezed like an old man dozing fitfully in his armchair. In desynchronizing their engines the Germans successfully defeated the British anti-aircraft defences whose guns were fitted with sound locators. They located only if the aircraft emitted a steady drone; if the Germans didn't, then the locators couldn't locate. But the eleven Heinkels had no need to worry about the AA defences as they approached their targets; they were flying 3,000ft above their maximum range on this night. What each of them scanned the silvery sky for were the British fighters. There were none above London as the first string of incendiaries clattered on to the east casements of the Tower of London and set light to the Constable Tower. Hundreds more incendiaries pelted down, lighting up the city with brilliant ribbons of fire. The Fireraisers turned south towards the coast and, God willing, Vannes airfield.*

The pilots of 74 Squadron had spent the early evening in the clubhouse of the Maidstone Flying Club on West Malling airfield.

* All eleven Heinkels of KGr100 returned safely, though one was damaged by an RAF nightfighter.

Outside, the ground crews gave each aircraft a pre-flight exam; they checked the oil tank was full, that there were 85 gallons of petrol in the tank, that the canopy was well lubricated and that there was nothing amiss in the cockpit. The armourer knelt on the grass under the wing and removed the panels that gave him access to each of the eight Browning machine guns. He slid in boxes of .303 rounds, 300 for each gun. The radio fitter opened the fuselage panels and inspected the crystals in the radio set, while the rigger fitted metal blinkers to the cowlings to blank out the flames from the exhaust during night flying, and tested the wheels and tyres. On such an icy night they warmed the engine regularly and the armourer covered the gun ports with canvas to prevent icing.

The pilots waited in dispersal. One or two were snuggled into their woolly fleece-lined Irvine jackets with their Mae Wests over the top. Most found them too much of an impediment, like Freeborn, who flew 'in my uniform and a lot of sweaters'. All of them killed time in the way the German bomber crews had killed time a few hours earlier. Dozing, reading, writing. Freeborn and Boulding shared a packet of Craven A cigarettes over a game of cards, 'for whatever money we could', recalls Freeborn. Whatever they were doing they all shared the same feelings of dread and diffidence. 'Don't believe anyone who says they weren't frightened,' says Freeborn. 'We all were and we all had ways of dealing with the fear. I used to tell myself that if I did get hit it would be a cannon shell in the head and I wouldn't know anything about it. Our greatest fear was fire.' Spitfires had a large armoured plate covering the fuel tank in front of the pilot but everyone knew someone who had been hideously burned by a cannon shell from a Messerschmitt. A 'flamer' was deep in the psyche of the boys of 74 Squadron. Mick Mannock, the squadron's greatest ace of the first war, had once sworn 'I'll put a bullet through my head if the machine catches fire ... they'll never burn me'. But the Germans did burn Mannock when they shot him down in flames in 1918. Such terror compelled a few Spitfire pilots to cover their faces with scarves and goggles so that no skin was left exposed. Others flew with the cockpit hood open so that they could bale out quickly. One or two carried a pistol.

At 11.15 the Spitfires of 74 Squadron were scrambled. Boulding and Freeborn rose to their feet without a fuss. The cards would have to wait. They joined the rest of the squadron in the short jog across the stiff blades of frozen grass that crunched under their weight.* Freeborn slipped his leather flying helmet over his head and grabbed his parachute hanging down from the wingtip. He struggled into it as the rigger removed the starter plug and pulled the trolley clear. The fitter, who'd already started up the Merlin engine, helped Freeborn into the cockpit and secured him in the harness. By the time the canopy was closed, the fitter and rigger were standing by the chocks.

Inside the cockpit Freeborn's anxieties had vanished. He felt 'part of his Spitfire', encased in 3ins of bulletproof windshield glass. Boulding felt equally at home in his Spitfire as he made his final cockpit inspection. He checked that the oxygen supply in the tube clipped to his helmet was working, that the R/T lead was plugged in, that the fuel gauges were full and the brake pressure fine. He waved 'chocks away' and released the brakes, eased open the throttle and started to taxi. For a few seconds his ground crew held on to the wingtips, then on Boulding's signal they let go. The Spitfire moved clumsily over the grass, lurching from side to side as bursts of throttle were tempered with a bit of brake. An ugly sight for such a beautiful aircraft, but the high nose of the Spitfire impaired forward vision so swinging her was the only way Boulding had a clear view in front of him.

Freeborn was already in position to take off. He opened the throttle

* This was and remains the coldest May night on record in England, with the temperature at Lynford, Suffolk, 75 miles north of West Malling, dropping to minus 9.4°C.

gently, checked he was in fine pitch and then the Spitfire started to bounce down West Malling's grass runway, guided only by the line of portable glim lamps. As the aircraft picked up speed there was still a bit of bounce. Freeborn brought the tail up a little as he felt the lift, the airflow over the elevators was perfect, more throttle, counter the swing with coarse rudder, more throttle. Eyes fixed on the instrument panel the whole time. Freeborn eased back the control column and up came the wheels.

Boulding trailed Freeborn up into the skies over Kent, climbing towards his patrol height of 18,000ft at 200mph in coarse pitch with slight movements of stick and throttle. Once he was airborne Boulding had the comfort of knowing that visibility for a Spitfire pilot was far superior to his adversary in a Messerschmitt fighter. There was headroom to crane the neck up, down and around and the blind spot was covered by the rearview mirror on top of the windscreen.

'There was radio silence this night,' says Freeborn, 'but each of us knew our job. It was a case of getting up there and seeing what you could do. But really for us [a day fighter squadron] it was more about luck than anything else.' Freeborn loathed night flying. 'We were pretty useless,' he reflects. Boulding considered the whole business 'not a very satisfactory proposition'. Their exhausts still gave off a nasty glare despite the blinkers and their presence actually impeded the pilot's view when he came in to land. Then there were the British anti-aircraft batteries. 'We used to get shot at by our own guns,' says Freeborn. 'They couldn't hit a German but they could hit us OK!' Freeborn had been winged by a British AA shell during an earlier night patrol and there was a tinge of trepidation as 74 Squadron approached London. 'We were to gain access to this central area via a gap in the [AA] barrage at 12,000ft,' remembered Boulding, 'marked by two coloured searchlight beams (blue, I think) fixed in the vertical position.' Once they were patrolling over the central area they would be out of range of the British guns, but climbing through the barrage was always hazardous. The two coloured searchlight beam idea was fine in theory, recalled Boulding, but in fact was 'almost entirely non-effective and I think most of us just flew through the barrage to our appointed patrol areas'. 'You knew if you'd had a near miss with an AA shell,' says Freeborn. 'You heard it and you saw the flash. When it happened you dived down to avoid any more.'

Now he was right over the central area Boulding was stunned by what he saw. 'London seemed to be at the base of a pyramid of flames – a truly horrifying sight.' Freeborn had seen 'London burning as I took off ... it wasn't nice at all'. He seethed silently. 'It made me so bloody angry,' he recalls. 'It was my country and they were dropping bombs on it. I felt very sorry for the innocent civilians and I wanted to do what I could to prevent it.'

ONE of the inconveniences of the Heinkel He111 was that there was no space for the navigator and his maps. Instead, Hufenreuter had to spread them out on his knees. At least tonight he could plot the course without use of a flashlight. Rarely could he remember a moon so bright. His disclosure of their new route hadn't been well received by the fretful Richard Furthmann. He had wanted to stick to the official route north over Belgium and Holland and into London along the Thames, but Hufenreuter overruled him.

So they flew over Cap Gris Nez, flashing their signal to the German AA batteries thousands of feet below. It was just the two of them in the cockpit now. Josef Berzbach had wormed his way into the underbelly gondola and was looking down at a sea that in the moonlight resembled a giant piece of tin foil. The only sound he and his four comrades could hear was the same uneven drone of desynchronized engines.

As they crossed the Channel, Hufenreuter took up position in the glass nose cone. The brightness of the sky alarmed him, and he ordered Furthmann to climb to 16,000ft as a precaution. They approached the English coast at 200mph, with Hastings on their left. From the Sussex town Hufenreuter navigated the Heinkel north-east towards Canterbury, as if taking his crew on a tour of historical English towns. Over Canterbury they swung due west to Croydon. Even before they reached the sprawling suburbs of London, Hufenreuter was able to see the ribbons of fires created by the Fireraisers of KGr100. He tested the bomb sight settings as a myriad searchlights swept the night sky seeking out the intruders.

Hufenreuter told Furthmann to lose height as they passed the Isle of Dogs. Puffs of charcoal smoke exploded around them as Furthmann levelled out at 9,000ft. He glanced down at the large watch strapped to his thigh. It was twenty minutes to midnight. Their allotted fifteen minutes to drop their bombs had expired. Over the intercom, Furthmann politely asked Hufenreuter if he would 'please hurry. Our time was up minutes ago'. Hufenreuter ignored the pleas and stared down through the bomb sight, his hand hovering above the push button on the trailing wire. Stepney was already burning well. 'Left, hard left,' he said, as if he were giving directions on an afternoon punt on the river. A large building caught his eye. A warehouse. His thumb pressed down on the button with precise calmness. 'Bomb gone.' The 1,000kg bomb hurtled towards Stepney with the electrical charge running through into the two spring-loaded plungers. As it fell at 500mph the charge went into a resistor, down into a condenser, into another resistor and stopped at the firing condenser. Seconds after leaving the Heinkel the 1,000kg bomb landed on Stepney, the impact of which activated a trembler switch, which made the contact, which fired the gaine, which detonated the bomb; by which time Hufenreuter and his crew were a mile to the west and preparing to drop four incendiary breadbaskets. They were released in quick succession, one, two, three, four dropping between London Bridge and Southwark Bridge. Cannon Street Station took the brunt.

Now it was time to get away from London. Their luck had held so far but the next few minutes would be the most perilous of the evening. They had bombed their target and survived the flak. They were alive. But crew's senses, fine-tuned on the outward leg by adrenaline, often became dulled by the relief that flooded their bodies. With the relief came tiredness, and the battle to maintain their concentration.

Hufenreuter rejoined Furthmann in the cockpit. He unfurled the map on his knees and began to chart the pinpoints that would lead them home: Maidstone, Hastings, Cap Gris Nez and Lille. He looked up at the altimeter, 17,000ft, and then at the luminous dials on his watch. 11.55.

ROGER Boulding had been in the air for thirty-five minutes when the problem started with the pitch control on his propeller. 'This wasn't too serious but made it advisable to cut short the trip.' He clicked on the radio, informed control at West Malling, and 'came down to approximately 17,000ft on a South Easterly course'. Apart from London burning he hadn't seen a thing the whole patrol. Suddenly in front of him, '200 yards to my starboard' he saw a twin engine machine heading in the same direction. He remembered Freeborn's words about night flying being more about luck than anything. 'I flew underneath and behind it and identified it as a hostile aircraft.' Boulding's hands gripped the 'stick', really a black wheel about 5ins across. If the wheel was a clock then the firing button was at 11 o'clock. Boulding turned the knurled ring of the button from the safety position. He looked through the sight, an oblong glass with a red circle and crossing lines. When he was 'between 50 and 100yds from astern and below at an angle of about 20°', Boulding pressed the fire button with his right thumb. The Spitfire shuddered as the eight Browning machine guns thumped into life.*

THE nose of the Spitfire dipped with the recoil and Boulding lost nearly 40mph of airspeed. The burst was a quick one, not more than three seconds, and Boulding 'saw De Wilde [ammunition] striking fuselage and a large mass of what looked like very large sparks came back at me as my burst reached the enemy aircraft'. Some metal part of the Heinkel had hit the Spitfire and 'damaged my air intake and grazed my airscrew'. The whiff of cordite filled Boulding's nostrils as 'the enemy aircraft immediately slowed up and dived steeply without altering course'.

Hufenreuter's scream of 'Dive!' was still reverberating round the cockpit as Furthmann thrust down the stick with his left hand. The first either had known about the Spitfire's presence was the red tracer that glided effortlessly past the port engine. Why the hell hadn't Berzbach or Weber spotted their stalker? Hufenreuter looked up at the instrument panel and saw only needles dropping. The port engine had been hit. 'Dive steeply,' he yelled. 'And fly bends.' Boulding trailed the Heinkel down towards Maidstone. He fired one short burst from port, then launched a two-second attack from starboard, but he now 'experienced some difficulty with icing-up on the inside of the bullet-proof windscreen'. He cursed as he tried to get the German in his sights once more. 'I had great difficulty in keeping behind the enemy aircraft,' he recalled. Furthmann was flying the bends ordered by his captain. He was also flying with the desperation of a man aware his grip on life was fragile. As the Heinkel trembled with the pressure of

* Each of the eight Brownings weighed 22lbs and had a muzzle velocity of 2,660ft per second. As the rate of fire of a Browning was 1,200 rounds a minute and they had only 300 rounds in each gun, they had to be conservative in firing. Most pilots had some tracer bullets at the bottom of the box to indicate they were about to run out of ammunition.

the dive, Furthmann fought the earth's gravity. The G force squeezed his body and his vision blurred as the blood circulation to his eyes decreased. Hufenreuter was experiencing the same sensations, but as more tracer glided past the starboard engine he ordered Furthmann to 'take her down as far as you dare'. He had no idea Boulding was firing blind but he knew they would only escape if they could get low enough to lose themselves against the dark background of the English countryside. Boulding knew it too, and pursued his prey with cool savagery.

Eggert Weber had missed Boulding's initial attack but now he was pouring fire at the Spitfire. He knew from the artful way the British pilot pursued them he was no novice. 'I weaved from side to side in an endeavour to keep track of him,' said Boulding, 'and noticed that when I was on the starboard side I was "up moon" of the enemy aircraft ... the top rear gunner could apparently see me as he fired some accurate bursts on that side but none on the port side.' Weber had fired a long burst at the Spitfire when it was silhouetted against the moon. Empty cartridge cases ricocheted off the Perspex dome as the British aircraft climbed to escape the moon. 'I got in two more short bursts at between 100 and 200 yards from port and starboard, slightly above and behind,' said Boulding, '[but] I had my work cut out to keep track of him through the glare of my exhausts.'

Hufenreuter was sure they had lost their attacker only when they were a couple of thousand feet from the ground and close to blacking out from the effects of the dive. Up above them Boulding circled the area like a prowling shark. When he'd obtained a radio fix he returned to base in his spluttering Spitfire and made his report to the intelligence officer. 'I last saw e/a [enemy aircraft] still going approx S.E. at about midnight,' he said. 'From my vector home to West Malling this would be about 10 miles S.E. of Maidstone.'

On board the Heinkel Hufenreuter had a choice to make, as the red warning light of the port wing winked at him in a panel above his head: crash land in England or try and limp across the Channel at its narrowest point on one engine. He chose the latter, ordering Furthmann to head towards Dover. It was an excruciating gamble; the Luftwaffe called the Channel 'der Bach' (the stream). Not many airmen survived a dip in its icy waters. Ditch in it tonight and Hufenreuter knew they were as good as dead. But their aircraft's inexorable loss of power mocked Hufenreuter's order to cross the Channel. The Heinkel lost height with a giddy sequence of shudders and clatters, until Josef Berzbach, lying behind his machine gun in the underbelly gondola, could see clearly the tapestry of the Kent countryside 1,000ft below. He deserted his gondola and crawled up into the catwalk, readying himself for the inevitable impact. Hufenreuter sat riveted in his seat looking straight ahead out of the glass nose. The frozen fields glittering in the moonlight reminded him of Christmas past at his home in the Harz Mountains.

As the aircraft roared low over some trees, Hufenreuter recoiled in terror. There were houses just in front of them. A church spire loomed up. They skimmed the rooftops with Furthmann rigid at the controls. 'Captain,' he said quietly, with no emotion, 'I can't hold her.' The aircraft lurched as Hufenreuter prepared for the impact. There was a jolt and he was thrown forward. The Heinkel bounced off the ice-metal ground and tobogganed with a demented shriek towards a thick row of hawthorn bushes. Beyond them was a line of centuries-old English oak trees. The aircraft ploughed into the bushes, ripping them out by their roots, and careered towards the oaks. Hufenreuter threw his hands up to protect his face. Furthmann was slumped unconscious over the control column. The Heinkel collided with the trees with a gentle thud. The starboard wing rested against the tallest of the oaks, the port one kissed a smaller, stooped tree. Snagged to the underside of fuselage were the hawthorn bushes that had broken the momentum of the aircraft.

Seventeen-year-old Peter Huckstepp had been woken by a 'roar of engines that seemed right overhead'. By the time he had leapt from his bed and pulled away the blackout curtain the plane had come down in the meadow opposite. Peter thought he had 'heard the sound of a man screaming' in the aircraft. He and his dad, Fred, an ambulanceman, tore downstairs. 'Grab your gun!' Fred yelled to his son. Clutching his Home Guard rifle against his pounding chest, Peter followed his dad across the road. They then turned left into a meadow known to locals as the Camp. 'The plane was about 100 yards away,' recalls Peter, 'and a bush had slowed it down and it had come to rest right up against another big tree.' The teenager drew back as the stench of aviation fuel hit him. 'I hesitated at that moment because I was scared the plane might blow up.' His dad had no such qualms and was already calling out to any survivors. The plane was painted black, he couldn't see whether it was friend or foe. But he could hear the soft senseless groans of men in agony. 'My dad was straight into the plane,' remembers Peter. 'The pilot was hanging half out of the cockpit but my dad managed to lift him clear.' As he joined his father the teenager glimpsed 'a big man walking round in a daze' underneath the branches of the small, stooped tree. It was too dark for him to see Hufenreuter's left lower leg and the milk-white bone that had punctured his flying boot. Hufenreuter asked Fred where they were. 'When he spoke in German my dad practically dropped the injured pilot. He'd lost two brothers on the Somme and he was no lover of Germans.'

Hufenreuter had come down a couple of hundred yards from Kennington, a small village just outside Ashford. Apart from the church and two pubs, the Rose Inn and the Golden Bowl, there wasn't much to it. But its inhabitants were well used to aerial warfare. During the Battle of Britain the Huckstepps, like most other villagers, had eaten their lunch in the garden so they didn't miss any of the dogfights. The latest excitement soon brought a bevy of curious neighbours. 'Edward Ward, the butcher, and his wife appeared,' recalls Peter, 'and old Mr Field who lived nearby. There was also an army camp just up the road so it wasn't long before soldiers arrived to take away the crew.' Eggert Weber whimpered in delirium as his crushed legs were prised from the swivel seat in the dorsal turret; Karol Gerhardt, the wireless operator, and Josef Berzbach were carried unconscious but alive into the ambulance. Albert Hufenreuter insisted on retrieving his forage cap from the cockpit before being helped into the ambulance. Someone wrapped an eiderdown round Richard Furthmann. Then they gently slid a stretcher under his broken body. A dozen hands lifted him into the ambulance, but he was dead by the time they reached the hospital.

13

RICHARD Furthmann's life had not been in vain. He'd played his small part in fanning the flames now taking hold of London. From West Ham to the West End, remembered one Luftwaffe pilot, the city was 'bubbling like a pot of boiling tomato soup'.

At 11.55 p.m., the time Furthmann's Heinkel dropped its bread-baskets of incendiaries on Cannon Street, driver Leslie Stainer of the Southern Railway was on his way to the station from the Bricklayer's Arms depot in Bermondsey to collect his train, the 12.53 a.m. Cannon Street to Dartford. It was a short journey but Stainer could feel the cold gnawing at his face as the engine nosed its way parallel to the Thames. 'A fire had started over by Surrey Docks and loads of incendiaries were dropped all the way to London Bridge and the City,' recalled Stainer. He braked just before London Bridge as a group of incendiaries dimpled the track. Harry Osborne, the engine fireman, jumped down on to the rails and trod them out. Then they pressed on. 'On arriving at Cannon Street, Platform 6, bombs began to drop,' said Stainer. 'A fire had then started at the side of the station, and it then rained bombs and there seemed to be no stopping. The fires were like huge torches and there were thousands of sparks.'

The station roof was now alight and the railwaymen's first thought was to save the trains. They coupled two engines together – Stainer with his driver's eye even remembered their numbers, '934 and 1541' – and pulled them out from under the roof on to the bridge. Ahead of them, 20yds away, was another train, stranded helplessly above the Thames. A Luftwaffe bomb aimer spotted them. 'We ducked down on the footplate,' recalled Stainer. 'We counted three bombs, the last one was terrific, and very close.' The titanic explosion scrambled everybody's senses and no one was sure what exactly happened in the maelstrom that swept over the bridge. Stainer remembered that 'debris flew in all directions'. Osborne yelled, 'Look out, we're going in the drink!' Through the thick choking clouds, Stainer saw that the 'bomb had made a direct hit on the boiler of No. 934 engine'. Like a giant kettle, the remains of 934 belched and hissed a steady plume of boiling steam. The bomb had also turned part of Stainer's train over on its side. Osborne tried to douse the flames taking hold with buckets of water, but it was no good. The breeze dancing west up the Thames soon fanned the flames. Stainer retreated from the bridge and looked to the Heavens to curse the bombers. He was aghast to see that 'smoke from the fires blacked out the moon'.

But by now the Germans no longer needed the moon to guide them, for London was banded black and gold. Some targets were easy to identify, like the Royal Albert Docks in West Ham, where a cluster of incendiary bombs fell with precise savagery at 11.50; others were easy to spot but harder to hit, like Southwark Bridge. The incendiaries meant for the bridge missed and landed on Southwark Fire Station at 11.51, damaging the chief officer's block.

And some buildings were just hit indiscriminately, like the Queen's Hall in Langham Place, just off Regent's Street, and the British Museum in Bloomsbury.

Just a few hours earlier the Queen's Hall had held 2,400 people listening in rapture to Malcom Sargent conduct the London Philharmonic Orchestra. Now, approaching midnight, the only voices were those of Tom Clark, the electrician, and Bob Rhodes, a fireman permanently stationed in the 21,000 sq ft of the Queen's Hall. The pair had just swept the building for incendiaries and found nothing amiss. They put up their feet in the caretaker's room just inside the artists' entrance in Riding House Street and filled the kettle with water. Suddenly they heard a heavy thud on the roof. They ran into the Hall and saw through the skylights sparks and flames coming from the side of one of the oval windows at the back centre of the ceiling, as though a workman was up there 'welding an acetylene lamp'. Rhodes thanked his good fortune. A 50ft hose was positioned just at that spot, and with the help of Clark the incendiaries were quickly extinguished. Clark told Rhodes he could turn the water off but before Rhodes had moved the hose went dry. 'It's turned itself off, Tom,' said Rhodes. Then there was a hiss, a sudden tremendous expulsion of energy and the flames were roaring from the incendiary once more. They men tried a hydrant at Block A in the balcony but it too was dry. Clark hurtled into the caretaker's room and phoned the fire service. They said they would be there as quickly as they could. Half an hour later flames were threshing back and forth across the entire roof of the Queen's Hall. When debris from the roof began to fall into the Hall there was still no sign of a fire engine; nor was there when the seats caught fire or when the blue-green paint started to wriggle down the walls. They hadn't appeared when the gilded pipes of the towering organ cracked and toppled, nor when the flames writhed their way underneath the platform into the band room where the instruments were stored. Inside they destroyed without discretion, devouring Amatis and Guarnerius and Stradivarius and cheap, worn instruments that lay beside them.

A few hours before the first bombs had landed on the British Museum, the museum's director, Sir John Forsdyke, a Greek scholar, small and implacable, had put the finishing touches to the biannual report he was about to present to the museum's trustees. It was an

burn for twenty minutes at temperatures of 1,300°C. The firewatchers put a foot through their pump to secure it in place against the bucket of water and started to pump water up through the tubing. As the timbered rafters disappeared beneath the flames the firewatchers pointed the nozzle at the roof. But they were intelligent men, all members of the museum's staff, and they appreciated the hopelessness of the situation, of trying to fight such an energetic fire with buckets of water and hand pumps. They withdrew and called the fire service. All Sir John could do was stand outside on the quadrant, watching and waiting. The first of eight fire service pumps arrived forty-five minutes after the call, by which time the roof of the Roman Britain Room was well alight. Flames were also licking at the general library, where inside hundreds of thousands of books were stored. And then Sir John thought of what else was imperilled: the Room of Greek and Roman Life, the Greek Bronze Room, the First Vase Room and the Prehistoric Room with its popular 'Suicide' Exhibition, outlining Europe's early history from 550,000 BC to the Norman Conquest. It had been promoted as the British Museum's 'sacrifice to the perils of war'.

Elsewhere in London, as Saturday became Sunday, most people had no idea of the gathering storm heading their way. The Fireraisers had come and gone, and twenty-two Junkers 88 of KG54 (Bomber Group) had vomited their bombs, but that had been the extent of it. Another raid, admittedly the first for three weeks, but nothing exceptional. Nothing out of the ordinary. Nothing to worry about.

From the doorstep of his house in Sydenham, on the southern outskirts of London, 72-year-old Reg Harpur and his cat, Peter, watched events unfold. It was so light he could even see several barrage balloons above the city. There was a big flash to his north, then an awesome red glare on the horizon. Harpur told Peter that was probably Camberwell Green getting it. Harpur's attention was drawn by a courting couple 'talking and laughing as they walked along the road but they took no notice of guns or shells or even that bomb, but went their way along Kirkdale as though as it was just an ordinary stroll late at night'.

Special Constable Ballard Berkeley was patrolling his beat when the first bombs dropped. He was standing outside the Lyons Corner House in Coventry Street talking in his actor's voice to the customers popping in and out of the restaurant, when a cataract of incendiaries fell from the sky at 250mph. They hit the road with their distinct and curious plop-plop sound and then erupted in a sizzle of bluish-white flame. Ballard watched 'helpless' with mirth as a man put a steel helmet over one of the incendiaries. 'The helmet went red hot, white hot and then disintegrated.' It had also amused the news vendor standing outside the Corner House with the evening edition stacked in front of him. 'Star! News! Standard!' he bellowed with a grin on his face. 'Star! News! Standard! Cup final result! Cup final result!'

'He just stood there,' recalled Ballard, 'and the bombs came down and he kept selling his papers.' Another bunch of incendiaries fell, just a few yards in front of a prostitute coming up from Piccadilly. 'She had an umbrella up,' said Ballard, 'and she was singing "I'm Singing in the Rain". The only rain coming down was the incendiary bombs. And I remember thinking ... I wish Hitler and Goring could have a look at this. It was quite extraordinary.'

In his flat on the Redriff Estate, young Tom Winter was still playing monopoly. Those already bankrupt had left the table and were watching the raid from the front door. 'Reddish glows could be seen looking in the direction of Tower Bridge and the City,' Tom recalls. 'So some other poor sod was copping it.' Just after midnight his dad popped his head round the door and briefed them on what he had seen so far. Bombs had fallen on Bermondsey and around London Bridge. There had been reports of casualties. 'He also said that so far as our particular neighbourhood was concerned, though it had been very noisy at times and a lot of activity seemed to be taking place up in the sky with aircraft dropping in and out of searchlight beams and gunfire banging away our area was reasonably free from serious incidents.' In fact, Tom's dad said with a wry smile, 'other than a couple of anti-aircraft shells that hadn't detonated up in the sky but returned to earth causing a little scare' it was quiet in their little corner of Bermondsey.

A good deal west of Bermondsey, across the other side of London, 21-year-old bombardier Bill Church of 164 Battery, Essex Regiment, was at his post on one of the four 3.7in AA guns in the park by Wormwood Scrubs Prison. Every time his gun fired one of its 28lb shells the windows on the surrounding houses rattled and the curtains danced a short jig. Encircled by the four AA guns was a command post with two spotters. 'They were past masters at recognising aircraft,' recalls Church, 'and they were always on duty peering through their telescope.'

AA gunners in London didn't have many friends during the Blitz. They were 'looked down on by the rest of the army' for leading what was perceived to be an easy life; the RAF loathed them for the eagerness to blast away regardless and civilians disliked them for duds and rattling windows and white hot shell fragments. On 10 May a warden was called to a house in Islington by an apoplectic householder. The warden's report noted laconically that an 'AA shell passed through the roof penetrating five floors and burying itself in the rear garden. Unexploded'.

There were few perks to being a gunner. By May 1941 most were suffering from hearing problems. 'We couldn't wear ear protectors,' says Church, 'because orders were given vocally. After a while the firing impacted your nerve ends in your ears.' And there were no cushy billets for the men. They slept by their guns, under blankets and stars, whatever the weather. During the exceptionally bitter winter of 1940/41 Church's gun had welcomed air raids. It got the blood moving and the red hot shell cases 'were put under our blankets to heat our bivouac'.

And they too suffered the capriciousness of their ammunition. 'Casualties from German bombs were light,' recalls Church. 'The most dangerous thing was shrapnel coming back down. You could sometimes hear it bouncing off your helmet, and there would also be quite a few duds coming down. And we never saw or heard them coming because there was so much noise going on.'

Tonight had been one of those cold, slow evenings. A Salvation Army canteen van had visited, 'a penny for a cup of a tea and a bun', and with the temperature below freezing the Salvationists handed out balaclava helmets and gloves to the men. There they sat, huddled round their gun with its cover off, ready and waiting. The 3.7in AA gun looked impressive. In the hands of well-trained crew its 4.7-metre gun barrel was capable of firing ten 28lb shells a minute, to a maximum ceiling of 25,000ft. Sound locators and the Fixed Azimuth system controlled the gun's fire. It was complicated, it was expensive and it was, recalls Church 'absolutely useless'. It would've worked, if only the German bombers had flown on a straight course and at a constant height and speed. But they didn't, and as the Azimuth worked only on constant sound, it couldn't locate the enemy planes with their desynchronized engines.

So when the Blitz on London had started the previous September, the AA guns remained silent, leaving the skies to the RAF, who were considered more likely to achieve success. But then the public clamoured to know why the guns weren't firing and it was decided, for the people's morale, to let them blaze merrily away in what General Frederick Pile, commander of Anti-Aircraft Command, later described as 'largely wild and uncontrolled shooting'. The civilians felt more protected, encouraged in their misconception by propaganda that liked to boast of numerous Luftwaffe bombers brought down by Ack-Ack fire. The gunners themselves knew the chance of hitting a German were slim, though they didn't couch it in quite the same mathematical terms as Professor Archibald Hill of the Air Defence Research Committee, who wrote a report on AA effectiveness: 'In order to give a 1/50th chance of bringing down an enemy moving at 250mph and crossing a vertical rectangle ten miles wide and four miles high about 3,000 3.7in shells would be required a second.' By the end of 1940 the GL (Gun Laying) Mark I radars had become available, so batteries no longer fired blind. Even so, the bearing and elevation of each enemy aircraft supplied to the gun crews by the radar was crude and inaccurate, and few planes were shot down.

'Take Post' was the cry that had brought Church and the rest of his nine-man crew to their stations just before 11 p.m. Gun layer, fuse setter, gun elevator, rammer, breach man, ammunition carrier, they all scrambled into position. Four minutes later Church's crew reported 'Ready for action'. For a brief, tantalizing, unnerving few minutes the only noise had been a harsh medley of plates, mugs and buckets banged against cell bars as prisoners protested at their confinement.* Then there came that old familiar uneven drone. The banging had stopped for a fleeting moment, and then begun again with a greater urgency. Then the noise was drowned out by the roar of Church's gun as the first shell shot out of the barrel and rocketed towards the moon at 2,600ft a second.

* The governor of Wormwood Scrubs Prison felt it was safer for everyone if prisoners remained in their cells during a raid. But Sidney Graves, a conscientious objector imprisoned in the Scrubs, recalled that 'during an air raid there was panic in the prison because we were all locked up'.

14

AT thirty minutes past midnight the storm broke over London. The smoke from the Fireraisers had dissipated and people on the ground looking up saw a sky lousy with German bombers. Thirty Junkers 88 from KG1, twenty-nine Junkers of KG77, fifty-nine Junkers of KG54, forty-two Heinkel He111s of KG55 and twenty-eight Heinkels of KG27, with a hundred more half an hour behind.

Bombs fell everywhere in those bedlam hours. They fell in the north, in Purcell Street, Islington, where an HE bomb flattened seventeen houses and left eight dead. They fell in the south, in Cunard Street, Southwark, where a landmine exploded on a row of houses owned by the R. White's Lemonade Company killing fourteen. They fell in the east, in Redmead Lane, Wapping, where a bomb landed on the premises of T. Allen Ltd, a cartage contractor, wrecking his ten horse-drawn vans, killing a driver and several horses. They fell in the west, in Notting Hill, where a covey of high explosives pulverized Bomore Road hewing out Nos. 12 to 40 on one side and 29 to 41 on the other. Seven civilians died and thirteen were wounded.

The bombs were paragons of democracy, paying no attention to wealth or social standing, religion or morality, history or heritage. In Buckingham Gate, two fire guards were killed as they fought a blaze at the Duchy of Cornwall offices. Six 500kg bombs screamed to earth in Greenwich, close to the National Maritime Museum, killing fifteen people and uprooting 300yds of roadway.

In Ebury Street the steeple of St Michael's Church, one of the few to escape the cull of steeples in 1940, crashed down on the head of 64-year-old fire guard Frank Gough. A bomb swooped down on the north end of Vauxhall Bridge Road. Two private hotels disappeared in the blast and the buildings either side were sliced open and exposed like dolls' houses. Both hotels were used by prostitutes to entertain clients and the proprietor of one was unable to tell firemen with any accuracy how many people were inside when the bomb hit. Thirteen bodies lay among the wreckage, some naked, others clad in fancy underwear.

For Dr Kenneth Sinclair-Loutit, making his way up City Road in Finsbury as head of a Heavy Rescue squad, the bombing put him in mind of what he had experienced during the Spanish Civil War. There he had led a British Medical Aid Unit to assist the International Brigade in 1936 and 1937 before returning to London to finish his studies at St Bartholomew's Hospital. He became a Labour councillor in Holborn, wrote articles warning of the dangers of Fascism and appeasement and counted among his acquaintances George Orwell, Stafford Cripps, Stephen Spender, Guy Burgess and Dylan Thomas, with whom he'd drunk pints in the Running Man pub off Bond Street. 'Thomas lacked conversational charm,' recalls Sinclair-Loutit, 'and Orwell obviously disliked me because our families had served in India during the Imperial period and I think he knew this which did nothing to generate camaraderie.'

At 12.30 a.m. on 11 May, Sinclair-Loutit knew he would have to draw once more on his sources of courage. Ahead of him he could see the Lipton's warehouse well ablaze. As he got nearer 'I scented an unbelievable and powerful aroma of good coffee'. Close to the warehouse were a 'number of crowded air raid shelters and I feared that such a blazing beacon could serve as a marker for the next wave of bombers'. Sinclair-Loutit bearded the fire officer and asked for a situation report. He was told the water mains had been hit and that they were fast running out of water. Sinclair-Loutit was by now squelching through a 'torrent of freshly infused good coffee'. Another fire engine arrived, a reinforcement from Chalfont St Giles, a genteel town in Buckinghamshire. 'Its volunteer crew were clearly, and with reason, worried,' recalls Sinclair-Loutit. A few minutes later they had used up the last of the water. The men stood helpless in front of the burning building. Then the doctor suggested opening a sewer cover and pumping back into the fire all the coffee. One of the firemen from Chalfont St Giles took a hose and, clinging to a ladder, was 'swung over to strike at the roots of the blaze ... this must have been the only example of fire-fighting with freshly infused café espresso'.

Sinclair-Loutit's attention was drawn to something else; a family emerging furtively from their side street slum. 'They were pushing a pram in which was their pathetically hidden family shame: an idiot child now virtually an adult, with a lolling head and a face showing a dimly comprehended terror. With exquisite tact the people in the nearby shelter made a space for them and rigged up a private corner screened with blankets.'

The Blitz had a way of forcing people to reveal their true nature. Character, like buildings, could be exposed by the emasculating bombs. On 10 May the Luftwaffe gave the men and women of London the opportunity to glimpse what lay within; not just in themselves but in others, too. Were they brave or cowardly, selfless or selfish, loyal or faithless?

At her home in Maida Vale at 12.30, Olive Jones was in bed trying to sleep through the raid when she was presented with her personal challenge. 'There was a tremendous loud howling whistle from a falling firebomb,' she wrote in her diary. Olive rushed to the window and saw that an incendiary had landed in next door's garden which was now 'full of a blinding, flickering saffron light'.

'So I ran out, shouting to the maids, and with some difficulty pulled myself over the wall by means of the branch of an elder tree at the bottom of the garden and ran to deal with it barehanded, fully expecting that one of the maids would come in a moment or two. But this was expecting altogether too much. Not one of the silly idiots stirred a finger to help me.' Using her bare hands Olivia covered the incendiary with soil and then found a pail with enough water to extinguish it. 'Just as I was getting through the last stages of my job a tremendous outburst of noise thundered from the sky. I ran like a hare to the wall ... I scraped myself over the wall somehow in a flap because of the noise as if the heavens were falling, but only to find myself like Absalom, inextricably entangled in the tree branches. I was still half-blinded by the brightness of the fire bomb and the awful blitz noise that filled the air seemed to wind mentally, for what seemed like hours I dithered and fumbled on the wall top, clawing frantically at the twigs that enlaced me and caught in my clothes.' Olive scampered back into her kitchen, where she 'found the maids blithering and giggling in the kitchen ... I gave them hell for their laziness'.

Gladys Shaw was feeling 'dusty all over ... I just wanted to take a bath, wash my hair and feel clean again'. She had been out on the streets for over an hour, visiting parishioners in the brick surface shelters. At first they had been good-humoured, she recalls, 'sipping their coffee and putting in their curlers and having a gossip'. Now they were scared. Gladys ducked into her next shelter, wincing at the reek of compressed unwashed flesh. She knew many of the parishioners but fear made their ashen faces anonymous. Now they were a trusting flock staring at her with questioning eyes. 'They loved it when I arrived,' says Gladys. 'It made them feel protected when I read from the Bible, as if they were receiving God's blessing.' Young women who had scorned religion a few hours ago now clasped their sweaty hands together. They lowered their heads, their eyes screwed shut, and joined Gladys in prayer.

Then Gladys moved around the shelter, stroking the hair of sniffling children and putting a hand of reassurance on the arms of those who sat in the corner trembling as if they were waiting to mount the gallows. She looked at her watch after twenty minutes and made for the exit with a mumble of departure. How she yearned to remain in the company of other people. 'I would have loved to have taken someone with me,' she says, 'because I was very frightened and I knew that when I stepped out on to the street I would be on my own. It was only seven or eight minutes at the most between shelters but that feels like an awfully long time when you're dodging bombs alone.'

Gladys left the shelter and headed south towards the railway line that ran west to Camberwell and then turned north to the Elephant & Castle. It felt as if there was a tornado coming up behind her. 'A piece of shrapnel came down behind me and I jumped out of my skin.' Gladys crouched down beside a wall too 'terrified to go on'. Up ahead was a railway arch. 'I looked at it and thought that if I go on there will be a direct hit just as I reach it. I began praying. I asked God to send a man to help me because I couldn't do it by myself. And then suddenly I regained my composure. "Don't be such a fool," I remember saying to myself. I got to my feet and ran under the arch as fast as I could until I reached the next shelter. I went in and someone said "Hey, look at her, looking so cheerful on a dreadful night like this!". If I looked cheerful I didn't feel it but I'd beaten my fear. The passage I read that night included the sentence, "ask what you will and God will give you his Holy Spirit" and I thought later that I'd asked for a man and there hadn't been one available but God at least had given me the guts to go on!'

Half a mile from Gladys Shaw's test of faith, John Fowler

staggered gingerly to his feet and looked around him. The scene of devastation seemed to have no end. 'I didn't know how long I'd been down for,' he says. 'I'd just started to walk home when all of a sudden I heard a big bang and that was it.' John checked himself for wounds and found his hair and face matted in blood. Dazed and petrified, he stumbled east along the Peckham Road, tripping over bricks and the detritus of people's homes, then he angled left into Southampton Street, right along Commercial Road, by which time 'fresh blood was streaming down me', and into East Surrey Grove. He knocked on the door of No. 111, his family's smart red-brick terrace house. His mum, Ethel, opened the door and lassoed her boy with two thankful arms. 'She cleaned me up,' recalls John. 'Although there was a lot of blood the wound wasn't that deep.' His mum asked what had happened, but John couldn't give an answer; he didn't know. He didn't know that Uncle Stan had been killed or that Cousin Rose was trapped under the remains of her house, bewildered but still clutching her pet dog. A parachute mine had exploded in Vestry Road demolishing his uncle's house and bowling John down the Peckham Road. John hadn't heard it coming, but few people did hear parachute mines as they wafted idly down to earth at the end of the green silk canopy. These 10ft envoys of the devil came in two sizes, 500kg or 1,000kg. If a parachute mine landed in water (they were originally naval mines) it became magnetic and exploded when a vessel passed over it. On land, however, they were high explosive bombs that because of their ambling descent rarely penetrated the surface but instead, exploded on impact, creating waves of destruction over a half-mile radius.*

'After my lucky escape my mum wanted me to go into their Anderson shelter,' says John. 'But I didn't fancy it so instead I made a bed under our kitchen table, which was big and thick, and stayed there.'

*Approximately 20 per cent of the 4,000 parachute mines dropped during the Blitz failed to explode because their fuses had not been adapted for use against land targets.

His parents returned to the shelter in the backyard. A mile and a half to the north-west the sky was a beautiful mix of red and orange. The poor old Elephant was taking it tonight.

THE Elephant & Castle was Superintendent George Adams's domain. It was part of 'F' District, which ran from Teddington to the Nore. At midnight as he watched the raid develop at the control room of his District HQ in Southwark Bridge Road, he saw call after call come in asking for more pumps to be sent to the Elephant. Adams told his driver, Chapman, to take him there so he could see for himself what was happening. Chapman turned left out of HQ and drove south down Southwark Bridge Road, underneath the Southern Railway line, and then turned right at the pub on the corner into Newington Causeway. It seemed at that moment they were the entering the fires of Hell. On either side of Newington Causeway shops and offices were burning, some insidiously, others with wild abandon. The colours were entrancing in their vividness: deep reds, bright oranges, the blue white of sizzling incendiaries, salmon pinks and rich yellows.

Chapman continued cautiously towards the road junction, into the heart of the conflagration. On their right was the Rockingham Arms pub and the Isaac Walton department store and on their left the Royal Albert pub. All three as yet untouched but the flames were coming up behind them. In front and around them was fire. New Kent Road, from Nos. 1 to 63 and 8 to 32, was ablaze. In Walworth Road incendiaries had dropped on Nos. 1 to 75 and 2 to 80. Newington Causeway and St George's Road had been clasped by fire and Nos 1 to 77 and 2 to 46 in Newington Butts were in flames. Of the Elephant's six starfish arms only London Road was clear. Adams ordered the handful of pumps already on the scene to each of the six arms. With each of these heavy pumps capable of pumping out 900 gallons of water a minute he told them to do what they could to contain the fire. Then he sent a despatch rider back to HQ requesting at least ten more pumps as 'an area of about five acres used as shops, dwellings, cinemas, churches, railway arches, warehouses, etc., severely damaged by explosion and fire.' He added one word, 'Conflagration'.

WITHIN a few minutes ten more pumps had arrived, just as an HE bomb landed in Draper Street, one of the side streets near the Elephant, wiping out a timber-strutted shelter under the Salvation Army slum post. Spurgeon's Tabernacle was alight by now, and Adams could see firemen chasing incendiaries on the roof of the Elephant & Castle pub. He didn't like the look either of the flames nuzzling the upper floor of Freeman, Hardy & Willis's boot shop or in the Park Press Printers in Newington Butts.

Most of the new pumps belonged to the Auxiliary Fire Service. They were trailer pumps, towed by a van or a London taxi, and much lighter than the heavy pumps of the regular fire service. They could pump out no more than 500 gallons a minute. Adams instructed the six heavy pumps to push back down their respective roads and establish a perimeter fence of water to stop the conflagration spreading. With more pumps arriving, Adams ordered the opening of the hydrants at the Elephant junction. These hydrants, fitted to the subterranean trunk mains, gave water at between 25 to 50lb per square inch [psi], depending on how many pumps were feeding into them. But this water pressure could be tripled by the pumps to as much as 150psi at the nozzle. The firemen opened up the first hydrant. It was dry. They opened the second, it was dry. The third, too, and the fourth. Every hydrant at the Elephant & Castle had run dry. Adams took the news phlegmatically; he still had other options. By Spurgeon's Tabernacle was a 5,000 gallon steel-panelled dam, one of hundreds set up earlier in the Blitz at strategic points in the city. The words 'Emergency Water Supply' (EWS) had been daubed on the side. He ordered one of the trailer pumps to feed into the dam and extinguish the fire taking hold of the Tabernacle. But he knew within ten minutes the dam would be emptied of its water. Less than half a mile away were the Manor Place Baths, where Joe Richardson liked to take his weekly ablutions, and which had been converted into another EWS. Adams directed two hose-laying lorries to the baths with orders to pump out steadily the 125,000 gallons of water through their 3½in rubber-lined hoses and back along the Walworth Road, across the Elephant junction, and into the steel dam. Three pumps were despatched the short distance up the London Road to the Surrey Music Hall at St George's Circus. Although London County Council officials had been incandescent with rage when they'd discovered a few months earlier that the basement of the bombed-out music hall had been cemented and turned into an EWS holding over 200,000 gallons of water, now this water might just save London's most important road junction. The three pumps drove right up to the music hall, which was illuminated by the flames shooting up from the Royal Eye Hospital. Seventeen firemen leapt into action. Except for 29-year-old Sub-Officer Edward Hollet, they were all Auxiliary Firemen. Most came from further out of London, from suburbs such as Mitcham, Forest Hill, New Cross and Merton. The one exception to the rule was Leading Auxiliary Fireman Jimmy Johnson. The 46-year-old was literally fighting to save his wife, Caroline, and the house he shared with her in Crampton Street, right next to the Manor Place Baths. In a few moments hundreds of gallons of water were being pumped down four 2½in hoses snaking back to the Elephant & Castle.

Back at the Elephant, Adams was having to shout to his sub-officers standing only feet away from him to make himself heard above the frothing hysterics of the air raid. Against a backdrop of the whining thunder of anti-aircraft guns, German bombs shrieked and whistled out of the sky and incendiary bombs plopped against the road and retched blue-white sparks. Already the air was alive with dust, the same peculiar yellowy dust Adams had come to recognize as the final exhalation of a dying building, and sparks flew around their heads like angry wasps. Dense clouds from high explosive bombs hovered in the air, smoke signals giving word of the devastation they'd just wrought. Reports reached Adams of water mains ripped open by the bombs: in the New Kent Road and the Old Kent Road and the 12in main in Newington Butts. The inexperienced firemen cursed as the ruptured mains haemorrhaged water. Where was the bloody turncock to shut off the valve and save the water? Veteran firefighters knew that the turncock had probably already turned off the valve but that the water would continue to surge from the bleeding main, taunting them, until it was dry.* With luck some of the water might collect in bomb craters, to be pumped out and played on fires. But most of the water evaporated in the searing heat of the conflagration closing in the Elephant & Castle.

Half a mile away the Reverend Christopher and Joan Veazey needed more water if they were to save St Mary's from incineration. They had been at home in their flat, recalls Joan, when 'someone phoned up and said "Do you know the church is on fire?" We shot straight over and could see the flames up on the roof'. In December an incendiary had lodged in the beamed roof and 'Christopher climbed 60ft up the side of the roof to put it out' but tonight the roof was already well alight. Christopher dashed inside and made for the stone staircase 'but the flames were too strong'. Forest FitzGeorge, St Mary's rector, who had won a Military Cross as a padre in the First World War, was with them but what could his courage do now? Still incendiaries 'rained down'. One of them suggested it might be worth trying to dislodge the bombs with sticks or stones, so in desperation they

* When a large trunk main – with a 36 or 48in diameter – was fractured it took a number of men over an hour to close down the huge valve, by which time the water had gushed out.

hunted around the church grounds for sticks and then hurled them at the roof, only to watch disconsolately as the sticks were consumed by the fire. Christopher picked up a rock and flung it skywards. It smacked against the roof with a clunk and then rolled down before plummeting to earth at their feet. 'We watched helplessly,' wrote Joan in her diary, 'our church gradually turning into a raging inferno.' The Veazeys stood inside their church, 'just a little out of the direct fire', mesmerized by the demonical scene. 'Each beam crashed down as it burned through.' Arthur Pullin, a local ARP warden, rushed past St Mary's on his way to another incident and marvelled at the effect the flames had on the church. 'The rose window in the east wall [was] filled as with glass of the brightest medieval orange.'

As the flames wrapped their tentacles around the altar, the Kentish ragstone of the outer walls growled as it expanded, then, as the heat became too much, the roof smashed to the ground with a roar of pain. The fire spread to the painted woods, adorned with figures of the Saints. 'They folded into grotesque shapes,' remembers Joan. 'It looked as if they were real people burning alive. We knew now that we would be the last people to worship in this huge church.' Frazzled pigeons dropped from the church tower and the pitiful cries of the birds still trapped riveted Joan to the spot. She looked up towards the heavens just as 'the great bells fell from their moorings and crashed in a mighty rush of sparks and red hot molten metal'.

BACK at the Elephant & Castle Adams's Divisional Officer, Geoffrey Blackstone, arrived to take control of the situation. He asked Adams for a situation report. He got one, terse and laconic. How many pumps he had, where they were, how far the conflagration had spread, the thirty broken water mains and the reliance on EWS. He also pointed out the scaffolding dam that been erected outside Burton's, the 50 shillings tailors, on the corner of St George's Road and London Road.

It could hold 5,000 gallons and was being fed by the EWS in the bowels of the Surrey Music Hall. Adams also informed Blackstone that a hose lorry was relaying water from the Thames at London Bridge, but that was one and a half miles away and it would take time. Blackstone sent a message to Divisional HQ. 'Make pumps fifty'. More firemen were needed at the Elephant.

More firewomen were needed, too. Women like Emily Macfarlane, whose fire station had been hit by an incendiary bomb early on in the raid. It had caused minimal damage to the roof of the Paragon School but it had caused a backwash of excitement for Emily, marooned at her telephone in the control room. The calls had started coming in not long after 11 p.m.; one of the first was from her own Peabody Estate, where an incendiary bomb had gutted a washroom on the fourth floor. Then at nineteen minutes past midnight she received a message concerning a huge bomb that had exploded in Munton Road. It was the same procedure as ever for Emily as she scribbled down the details on the standard message form. Date, time and her initials along the top line. Below that 'Position of Occurrence'. Then 'Type of Bombs', with 'HE/Incendiary/Poison Gas' the three options. Further down there were spaces to jot down details of roads blocked, damage to mains, casualties, position of bombs and services already on the spot. At the very bottom was a large empty box, with 'Remarks' written in the top left hand corner. Most of the time Emily just wrote 'Fire Service Reqd Immed' and handed the message to the Mobilising Officer. She didn't think about how much pain and suffering and grief might be contained in that one thin slip of paper.

When Blackstone requested fifty pumps to the Elephant, it was a cry of 'all hands on deck'. Emily was told to take the canteen van and dish out rations to the firemen. 'I couldn't cook myself, I hated cooking, but I could drive and not many women could,' says Emily. There was a cook at the station, an older woman, whose speciality was shepherd's pie. Emily, the cook and a couple of other firewomen began driving up the New Kent Road towards the Elephant. 'It was so bright you could read a paper,' recalls Emily. Halfway along the New Kent Road, Emily drove past the tenement flats in Munton Road, or what remained of them. Twenty-four people were dead under the rubble, including the Keefes of No. 261, Edward and Lucy Keefe and their five children: Lilian 11, Lucy 9, Alfred 5, Anne 4 and 2-year-old Tommy. Over Chatham Street one of those compact clouds of peculiar yellow dust lingered over what was left of Nos. 53 to 61. The Smiths, Collins, Vincents and Turners, all obliterated by a direct hit. Emily's air raid warden brother had been nearby, she later learned, and the blast sucked out his lapel badge and deposited it 100yds away, where a comrade chanced upon it the next day and handed it back.

The top end of the New Kent Road was impassable, so they parked the canteen van under the railway bridge of the Southern Railway. A bomb had already hit the bridge so they reckoned that the law of averages favoured them. 'The heat was just so great,' recalls Emily. 'It was unimaginable. Cook started dishing out rations to the men and I did what I could to help.' The heat parched throats and the rolling clouds of dust made talking painful. It impregnated their clothes, their hair, their nostrils. The dust seeped into every pore of their bodies.

By 1 a.m. word reached Blackstone and Adams that the Manor Place Baths had been sucked dry of water. The Surrey Music Hall basement was still pumping a reassuring flow of water, and Emily remembers that 'a bombed sweet factory just by my Peabody estate that had been turned into a water reservoir' was providing a further source. There was also the relay from the Thames at London Bridge. The two officers discussed whether to open a second relay from the river, between Waterloo and Westminster, but decided to first take water from the Surrey Canal in Camberwell.

Outside St Mary's Church Joan and Christopher Veazey were

numbed with dejection and fatigue. For the best part of an hour they'd fought a gallant battle to save their church, scooping buckets of water from the small tank in the basement of the rectory and throwing them on the flames. 'It was like pouring a thimble of water on to hell itself,' says Joan. 'My feeble efforts were no use and by now my baby was jumping like a salmon going up river to spawn.' Their efforts hadn't been totally in vain. A heap of religious artefacts stood like a cairn marking the spot of a fallen comrade; silver chalices, registers, a few kneelers and a dozen prayer books. They'd forgotten the vestments and it had been 'impossible to save the altar frontals' but not all was lost. 'When Christopher realised that we'd been beaten he said "we had better go and see if someone else wants help". So very wearily – I would not let him out of my sight, baby or not – we walked up towards the Elephant & Castle.'

15

AT Waterloo Station the gents' lavatory was chock-full of stranded rail passengers lying on its black and white marble floor. Trains had been temporarily suspended and the safest place in the station was the cavernous basement lavatory, 80ft by 40ft, with its own boot cleaning room and hairdressing saloon. When the raid started Jean Ratcliffe and her boyfriend, Clement Edwards, had been meandering back towards Waterloo having watched Vera Lynn and Max Miller at the London Palladium. Clement was about to go into the merchant navy as a wireless operator and this was his adieu to the girl he'd been courting for two years. Still on the northern side of the river, the 21-year-old lovers had been 'pushed into a tube station by an air raid warden'. They had descended to the platform but the smell of fetid bodies and stale urine overpowered them. 'We escaped and ran over Waterloo Bridge to the station hoping to get back to Epsom.' Not long afterwards a high explosive bomb tore a 15ft crater in the roadway, rendering the bridge impassable just a week after it had been reopened. 'Once we got to Waterloo,' says Jean, 'we were ushered down the gents' loo. I was a bit prissy in those days and when I got halfway down the stairs I refused to go any further.' Clements and Jean sat down on the steps and were soon joined by a couple of army officers. 'It was a very chatty atmosphere and everyone was very nonchalant,' remembers Jean.

At half past midnight a German bomber had flown up the Thames and dropped a ribbon of incendiaries on the southern bank from Waterloo up as far as Lambeth Palace. Six minutes later a second bomber had followed on the same bomb run and used the fire coming from the London Waste Paper Company in Belvedere Road, running parallel to Waterloo, as a marker. A high explosive whistled down and landed in the bonded warehouses of York Road, which were in railway arches on the north side of the station. There were a lot of goods stored in these warehouses, including the Savoy Hotel's most valuable silver, glass and china. Fred Cockett recalls in the vaults that there was also a great deal of 'White Horse whisky and other spirits' valued at £30,000. The alcohol was stored in giant tuns and the door of each warehouse wore three padlocks, belonging to the owner of the spirits, the owner of the vault and the inland revenue. At 0036, the London Fire Service HQ received a terse message from Waterloo: 'Eight railway arches about 650ft used as bonded warehouses, offices and stores severely damaged.'

The blast from the high explosive had disembowelled the tuns, sending gallons of alcohol cascading on to the floor of the railway arches. More incendiary bombs were dropped and by 1 a.m. the arches were flaming like a colossal Christmas pudding. None of this was of concern to Fred Cockett and the firemen of 101 Waterloo Fire Station. They already had a serious fire on their hands. 'We were desperately trying to save our bakers!' he recalls.

The first bomb to hit The Cut landed on H.W. Sexton, the haberdasher, at No. 18. Four minutes later a high explosive destroyed Bronstein's the tailor, Boyd the grocer and Verrico the barber, where Fred and the boys liked to have a trim. Two minutes later Wood's Eel and Pie shop, and Raida the draper were reduced to a tapestry of bricks, glass and mortar, and a minute after Nos. 104 to 116 disappeared. The linoleum warehouse was hit at 1.35 a.m., the same time as Plander the baker, and at 1.40 a smatter of incendiaries landed on Nos. 62 to 78. Crisp the baker, at No. 74, was where Fred bought the station bread each morning and the thought of losing Crisp's spurred the men into redoubling their efforts. 'We knew we had to stop the fire spreading down The Cut,' recalls Fred, 'so we decided to stop it by our baker's. The problem we had was no water because bombs had fractured the 24in main in Waterloo Rd and the 12in main in York Road. So I got onto [fire service] control and it was somebody I knew so I asked for assistance. "Fred," he said, "brigade workshops are alight and I haven't got a pump for them so what chance do you stand?"'

Fred remembered a recent conversation he'd had with a 'pal of mine who'd transferred to a station north of the river near Soho ... they had five hose-laying lorries but not enough crews for them'. Cockett decided to take a gamble on there being at least one of the lorries free. 'I got the staff car and told one of the drivers to take me there. Off we went, over Westminster Bridge, past the House of Commons, which was alight, and then right towards central London.' They arrived at the station and all the regular pumps had been called out; a couple of the hose-laying lorries, however, were still there, key in the ignition, waiting in reserve. 'I didn't go into the ward room to ask,' says Fred. 'I just climbed into the first machine, started it up, and drove back to Waterloo'. Coming back over Westminster Bridge, Fred drove straight and whispered a little prayer. 'I couldn't see anything because of the smoke and darkness.'

Fred drove the lorry down to the Thames at Canterbury Wharf, on the southern end of Waterloo Bridge, where a crew from Kent* was seeking gainful employment. 'The hoses were the standard 2½in size so we connected them to the Kent pump and then I synchronized my watch with the fireman in charge. I told him to start pumping from

* By 1 a.m. all of the London Fire Service's 1,270 pumps were in action and the 1,242 pumps from outer London and the home counties began moving into the capital to reinforce their London colleagues.

the river in ten minutes.' Fred raced back to The Cut where his comrades were struggling to contain the fire. 'We quickly erected a mobile dam, which the lorries carried,' says Fred. 'They were canvas with a metal framework and it only took three or four minutes to put one of them up. They could hold 1,000 gallons.' The Kent crew were true to their word. After ten minutes the nozzles of the flat hoses trailing from the river along Waterloo Road into The Cut started to whistle like a kettle as the air was driven out by the water. Then the hoses writhed violently for a second and went taut as the dirty water of the Thames started surging into the dam. As quickly as it came in it was pumped back out and on to the fires spreading remorselessly along The Cut. The firemen fought the fires in pairs, crouched low over the hose with a solid stance, gripping its ice-cold nozzle, and with every sinew pushing against the backward thrust of the water pressure. The water that shot out of the 1in nozzle looked like an endless iron bar. 'Within five minutes we were soaked to the skin from the back spray,' recalls Fred, 'which was terrible on a night as cold as 10 May.'

Official fire service policy, as laid down by some bureaucrat in the Food Ministry, was to use only fresh water from the mains on food warehouses and grocery stores so as not to contaminate the food. Only in emergencies was water from rivers or canals to be used. A worthy theory but absurdly impractical, especially on a night like tonight when everything in The Cut, from Crisp the baker to Cox the butcher, was hosed in the water from the Thames.

And all the time the bombs kept coming from the Heinkels, tumbling from the nose-up position of the bomb bay to the nose-down position as they whistled and screamed through the air. 'Most of the time we were so engrossed in the fires you didn't have time to worry about the bombs,' says Fred. One high explosive landed on the southern side of The Cut, just up from the Old Vic Theatre, blowing apart Nos. 83 to 101. The tidy terraced two-storey houses of Roupell

Street, running adjacent to The Cut, were tattooed with several dozen incendiaries. Across the road from Roupell Street, in the warehouses underneath Waterloo Station, the fires were cavorting out of control. The heat was so intense that the platforms above were bubbling and the fumes from the alcohol tuns were so potent that the first crews on the scene were now intoxicated. Pumps from other parts of London and its suburbs were called in to take over.

One of the firemen who sprung into action was 29-year-old Glaswegian William Young, the driver of an auxiliary pump at a sub-station in Winsland Street, Paddington. Young had square shoulders and a thin petulant face and he was brave and dogmatic. His Cockney pals called him 'Mac', a term of affection because no one took liberties with him. Mac had been the assistant manager of a London hotel in the late 1930s and when war was declared he joined the fire service instead of the army so he could be near his sick wife. 'I thought the fire service would be fun,' he says, 'ringing bells and all that.' On Friday 1 September 1939, the day the Civil Defence was mobilized, Mac was one of the firemen who went to Shepherd's Bush taxi depot and commandeered their fleet of taxis for the Auxiliary Fire Service. 'We put a drawbar on the back of the taxis and used them to tow our pumps.' It was carried out, as far as Mac remembers, with little opposition from the taxi drivers. There might have been on or two scuffles but most drivers saw it as doing their bit for King and Country. Quite a number joined the AFS as drivers, helping Mac and the other drivers learn London's short cuts.

Throughout the Phoney War and into the Blitz, Mac had looked after his pump with the same tenderness he'd shown for his wife. He took it for regular check-ups at a garage in the Marylebone Road where 'they tuned my trailer pump engine so it was like a racing car engine' and he slept within 15ft of it in a canvas stretch 6ft by 2ft. When his station was called out Mac was dozing. 'The turn-out time for my pump was 18 seconds. I just threw off my blankets, jumped into my leather boots, pulled up my leggings and clipped them on, put on my helmet and then started up my pump.' Two firemen jumped in the back and a third hopped into the front passenger seat. 'The duty firewoman appeared and told us "York Road". I said, "Where the blink is York Road?" She went to query it, came back and repeated, "York Road". I said, "Well, the only York Road I know is down by Waterloo Station and that's way off our ground". "That's the one," she said. So off we went, ting-a-linging to York Road.'

Mac drove down Edgware Road, round Marble Arch, along Park Lane, past Victoria Station, where police and firemen were running helter-skelter across the forecourt, and over Westminster Bridge, lit up by the flames shooting up from the Houses of Parliament. 'When we got to the entrance to Waterloo the fire underneath the station was terrible,' recalls Mac. He had just parked the pump by the steps at the end of York Road when a 'string of bombs came down, I seem to remember three'. Everyone in London during the Blitz had their own technique for surviving bombs by which they swore. Some dodged into doorways, others threw themselves flat and hugged the ground. A few remained upright just so long as they could hear the bomb's whistle. Mac, like all firemen, had been taught that the blast from a high explosive had an outward and a suction wave and that, like sound, blast travelled in all directions. Caught in the open by an HE, a fireman was told to lie flat on his stomach with his chest raised off the ground to protect his ribs being crushed by the vibration of the detonating bomb. The explosion from an HE, if you were near enough to it, felt like there was a troll far below hammering against the surface with a giant sledgehammer. Mac counted the explosions: 'One, two, three and if the noise got louder with each one you knew the next was coming your way.' He threw himself on to York Road, 'chest slightly off the ground with my head towards the sound of the bomb because shards of glass waved over us like leaves. I pulled my steel helmet over my face to protect my eyes'.

The bomb exploded about 40yds from Mac and his pump. They waited until the troll had stopped hammering and then clambered to their feet. 'The dust had hardly settled when this WVS [Women's Voluntary Service] lady came along pushing a tea trolley, so we all had a good cup of tea.'

A fire officer arrived and directed Mac's pump to one of the offices on York Road. 'There was a pair of huge wrought iron gates padlocked,' recalls Mac. 'We all looked at each other as if to say, "How do we get in there?" One by one the five men climbed the gates and dropped down, trailing the hose with them as they pushed further towards the seat of the fire. 'With every step of our leather boots plaster was falling from the roof of this place,' says Mac. 'The fire was in one of the back offices and had clearly been going for some time when we arrived. There was a big long table, a beautiful-looking committee room table, and it was well alight. The officer turned to me and said "What should we do?" "Make pumps ten?" I replied.'

16

WHEN Mac Young had driven over Westminster Bridge it seemed to him that London was 'doomed'. There was nothing but fire. Nineteen-year-old Kathleen Abbot was standing on Hammersmith Bridge with a friend on her way home to Barnes. 'It looked like a painted sunset,' she recalls. 'It was selfish of us, I know, not to consider that people were dying but we just thought how beautiful the flames looked.' The underside of the silver barrage balloons over the city reflected the flames and added to the panoply of colour. Soon the moon changed its hue, from silver to a dirty yellow.

The first incendiaries to hit St Thomas's Hospital had spattered Riddell House at eleven minutes past midnight, from where a few hours earlier the Archbishop of Canterbury had given 'an inspiring address'. There was the familiar series of short blasts on a firewatcher's whistle, the alert for falling incendiaries, and men began running towards them with stirrup pumps. At sixteen minutes past midnight, Block 4 was hit and the roof set alight. Annie Beale, the Sister Casualty, began receiving the wounded as 'incendiary bombs fairly whistled and rained down on us without ceasing ... we in Casualty felt that the roof must crash down on us at any moment'. More incendiaries plopped down, on Riddell House again, and in the timber store. Three 50kg HE bombs hit the main building and one dropped in the hospital garage, grievously wounding Evan Morgan Jones and Robert Tanner, two auxiliary firemen. They were brought into Casualty, which had moved into the basement because of the 'terrific noise', but died amid the uproar. Sister Beale and her nurses turned to the living, plucking shards of glass from firemen's faces or tending eyes blinded by the sparks from the winds whipped up by the fire's air currents. She received regular updates from the firemen, all of whom told of the 'difficulty of procuring sufficient water with which to fight the fires'.

Water was making men mean in the early hours of 11 May. Station Officer Charles Davis was husbanding his scant resources as best he could at St Thomas's when an out-of-breath verger accosted him. 'Lambeth Palace roof is on fire,' he blurted out. 'We need help'. Davis told him he had 'no engines to spare'. 'But the Archbishop of Canterbury is in residence', the verger insisted. 'I'm sorry,' said Davis, 'but my priority is the hospital.'

A few minutes later Davis successfully signalled with a lamp to a fire float on the Thames and the fires in St Thomas's were gradually brought under control by their hoses.

For Cosmo Lang, the Archbishop of Canterbury, there was no divine inspiration. A stick of four HE incendiaries fell from a Heinkel on to Lambeth Palace during that hellish half hour after midnight. The first exploded in the Lollards' courtyard, the second in the main courtyard, the third on the wall between the churchyard and the courtyard and the fourth on the chauffeur's house.

Cosmo Lang wrote after that he was 'quite near' the bomb that exploded in the Lollards' courtyard. 'The blast induced me to fall flat! Fortunately a brick blast wall had just been erected in front of the door where I happened to be, which took most of the blast. Otherwise I might not be alive now.' Lang picked himself up and insisted on helping chaplain Ian White Thomason with the evacuation to Lambeth Bridge House of the two hundred parishioners sheltering in the crypt under the thirteenth-century chapel. It was, recalled White Thomason, 'an unpleasant though successful operation', and completed just before a flight of incendiaries descended on the Palace. Within minutes three fires were raging unchecked on the roofs of the library, the chapel and the Lollards' Tower.

White Thomason screamed inwardly at the impotency of the fire service, accusing them later of being 'hopelessly inadequate to deal with the situation … it was a long time before an engine appeared'. He and other members of the Palace staff, 'assisted by a number of boys who'd been using the crypt as a shelter did all they could to fight fires but there were not enough of them to deal adequately with three large simultaneous holocausts'. Throughout this trying time White Thomason was struck by the 76-year-old Archbishop's 'complete indifference to danger'. Perhaps Cosmo Lang's detachment was quiet fulmination as he reflected on a string of correspondence he'd had with the Civil Defence authorities. Two months earlier he'd requested a 'water supply and hose for the chapel roof … 20 stirrup pumps, ladders, 40 buckets and also [water] tanks and sandbags'. None of these requests had been granted on 6 May, prompting the Archbishop to issue a fierce rebuke to the Civil Defence. The next day Cosmo Lang was told to be patient, the items would soon arrive; but they hadn't by 1 a.m. on 11 May, by which time flames were jumping out of the top two floors of the Lollards' Tower, where heretics had been imprisoned in the seventeenth century, the west end of the chapel roof had burnt through and half the library roof was on fire.

Directly across the Thames from Lambeth Palace in Westminster, Canon Alan Don was unaware of the plight of Archbishop Lang as he slipped out of 20 Dean's Yard to check on St Margaret's. It had been an hour since the siren sounded and in that time his little corner of Westminster had remained bomb free. The explosions and fire bells came from south of the river. As Canon Don was shutting the front door of his house, Margot Seymour-Price was making her way across Little Dean's Yard from 'A' Division control room to their kitchen in Westminster School. 'We thought it would be a good idea to have a mug of tea,' says Reenie Carter, 'so Margot went across the yard and up the stairs to the kitchen. Unfortunately as she waited for the kettle to boil the Germans arrived and when Margot came down the stairs with a tray of tea she found there were incendiaries everywhere. We could see her standing in the doorway but she couldn't get across until there was a break in the bombing.'

Don reached St Margaret's and ducked down behind some sandbags by the north transept door just as 'a shower of incendiary bombs dropped all round on the [Westminster] Abbey roof and in various parts of the precincts and on the grass around St Margaret's'. One landed outside the vestry opposite Westminster Hall and started spitting out a thousand sparks. Don seized a sandbag and clubbed the bomb to death. Glancing up he saw a 'considerable fire was spreading in the offices below the Hall. Several fire bombs penetrated the roof of the Abbey and lodged in the Triforium where they were dealt with by the Abbey firemen'. But the incendiaries outnumbered the firemen. One set light to the timbered roof on the north side of the Abbey above the tomb of the Unknown Warrior. On the southern side of the Abbey, in the Little Cloisters, the seventeenth-century house occupied by Canon Michael Barry was ignited by incendiaries. Armed with a stirrup pump, Barry, wearing his blue serge ARP warden's uniform, attacked the flames now lapping his house, but within the hour it was submerged under the fire.

Margot Seymour-Price had made it back across Little Dean's Yard, dodging and weaving with the tea tray gripped tightly in her small hands. Grinning broadly, her colleagues complained that the 'tea was a bit on the cold side'. Then a shoal of incendiaries fell behind her on the school, illuminating the roof like fairy lights on a Christmas tree.

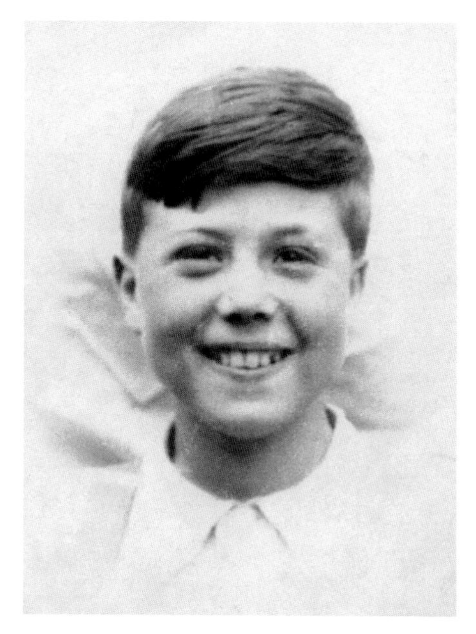

TOP LEFT Emily Macfarlane (left) grew up in the Elephant & Castle and witnessed its destruction on the night of 10 May 1941.

TOP RIGHT John Fowler and his cousin, Rose, got caught up in the bombing as they returned from the Peckham Odeon.

LEFT Bermondsey girl Gladys Jenner was on her way to the shelter at the Peek Freans premises when it took a direct hit.

ABOVE After the Blitz Joe Richardson joined the navy and later became a London taxi-driver, but the memories of 10 May continued to haunt him.

ABOVE Minutes after this photo was taken on Joan Veazey's wedding day in September 1940 a German aircraft swooped down and attacked the guests.

TOP RIGHT At her Sunday School class on 11 May Gladys Shaw asked the children to pray for the German airmen who had just bombed London.

CENTRE RIGHT Bombs destroyed the Stepney house of firewoman Florrie Jenkins on 10 May as she worked round the corner in the control room of the local station.

RIGHT Anne Spooner was a former debutante who joined the ambulance service because she wanted to do her bit for Britain.

ABOVE LEFT During the Phoney War of early 1940 William 'Mac' Young (front row, centre) and his fellow firefighters were called 'army dodgers' by the public.

ABOVE Each day Fred Cockett kissed his family goodbye in Brockley and cycled into work at Waterloo Fire Station.

LEFT Life in the early 1930s was more fun for Lew White (left) than it was during the Blitz when he was an Auxiliary Fireman in Holborn.

TOP Trailer pumps in Farringdon Street suck out water from one of the 5,000-gallon steel dams maintained by relay from the Thames. The fact that petrol often leaked from the pumps doesn't seem to worry the fireman enjoying a fag (second from right). (IWM MH15549)

ABOVE A trailer pump, similar to the ones Bobbie Tanner replenished with petrol with such nerveless courage on 10 May at the Peek Freans factory. (IWM HU1175)

TOP The firewomen of A Division control room pose in the garden of Westminster Abbey with Reenie Carter second from left and Margot Seymour-Price on the far right.

LEFT Bobbie Tanner, the only London firewoman to win the George Medal in the Second World War, with her uncle on the day she received her award.

ABOVE Firefighters reunited: Among this group of veterans posing by a fire engine in 2004 are Fred Cockett (second from left), Lew White (second from right) and Reenie Carter (far right).

ABOVE Roger Boulding cuddles his dog, Sam, as Henry Szczesny and Johnny Freeborn (right) kill time in the dispersal hut with a game of cards.

RIGHT Guy Gibson (right) shakes hands with Richard James underneath the nose of their Beaufighter, Admiral Foo-Bang III.

OPPOSITE TOP Albert Hufenreuter (second from right) in a POW camp in Canada in 1943.

OPPOSITE RIGHT Though not taken in May 1941, this Luftwaffe reconnaissance photo shows the Isle of Dogs in the U-bend of the Thames. The Redriff Estate and Surrey Commercial docks are under the starboard wing. (IWM C5422)

TOP No. 35 Bomb Disposal Section, Royal Engineers, in 1941, with Harry Beckingham on the far right of the front row.

ABOVE One of Harry's comrades stands next to a defused 1,000kg 'Hermann' (the sort that dropped on Ballard Berkeley in Soho), while cradling a 50kg bomb.

OPPOSITE TOP The end of the Elephant & Castle as flames engulf another office block. (London Fire Brigade)

OPPOSITE RIGHT A woman is brought to the surface by a London rescue squad, men famed as much for their irreverence as for their courage. (Popperfoto)

R
R

ABOVE Looking down into the Commons Chamber and the bottle-shaped hole on the far wall is all that remains of the doors that lead to the Members Lobby. (Getty Images/David/Stringer)

OPPOSITE TOP The Dean of Westminster, Paul de Labillière, stands in a pool of water before the high altar and sifts through the flotsam. (Getty Images/Keystone/Stringer)

OPPOSITE LEFT All Sir Henry Wood could salvage from the Queen's Hall in the days after 10 May was a cymbal. (Getty Images/Fox Photos/Stringer)

TOP For the first 50 seconds an incendiary bomb burned at 2,500°C before fizzing for a further 20 minutes at temperatures of 1,300°C. (Note the trees ringed with white paint on the London pavement behind, for guiding pedestrians during the blackout.) (IWM HU662)

CENTRE Fires run amok on the New Kent Road in the early hours of 11 May. (London Fire Brigade)

RIGHT Ballard Berkeley was blown off his feet when a high explosive landed here in Old Compton Street in Soho. (Westminster City Archive)

TOP Marylebone was the only mainline station not hit during the raid. The rest, like St Pancras, above, suffered varying degrees of damage. (IWM HU667)

ABOVE Londoners step over hoses on their way into work on Monday 12 May as firemen continue to fight fires in Farringdon Street and Great Bride Street. (IWM HU644)

ABOVE Looking down on London from the top of St Paul's Cathedral a few hours after the city's worst raid of the war. (London Fire Brigade)

RIGHT The towering inferno of St Clement Danes in the early hours of 11 May. (St Clement Danes)

OPPOSITE Ludgate Circus, with St Paul's Cathedral in the background, chokes with smoke from the fires still burning on Sunday. (IWM HU36221)

SMITHS'
FLOORS
FARMER & SONS
PUNCH
TAVERN
LEW

LEFT Flames roar out of a building on Queen Victoria Street, just a few metres from Faraday House, the telephone exchange that kept Britain in contact with the outside world. (IWM HU649)

BELOW No. 23 Queen Victoria Street comes crashing down on 11 May. (IWM HU650)

In the next few minutes Reenie Carter, Margot Seymour-Price and the other firewomen in the control room compiled a blow by blow record of the Palace of Westminster's disintegration. Its terseness was its eloquence, a testament to the professional detachment of the young women calmly taking calls in their pitiful brick control room.

> **0028: The Sanctuary, Westminster Abbey**. 'Hit by incendiary bombs [i/bs]. Dean's residence building of 3 floors about 100ft by 50 used as dwelling and office. About two thirds and contents severely damaged by fire and most of roof off.'
>
> **0036: Palace of Westminster**. 'Ministry of works and buildings hit by i/b 'About 80 x 40ft of roof damaged by fire. In the Royal Court back room on 4th floor and contents damaged by fire heat and smoke and water. Ceilings and contents (including King's Robing Room) under water.'
>
> **0036: Westminster Hall**. 'About 80 x 40ft of roof damaged by fire. Empire parliamentary officers (annexe) severely damaged by fire and most part of roof off, ceiling and contents under by water.'
>
> **0036: House of Commons**. 'A building of two floors about 100 x 60 feet (used as assembly hall and offices) contents severely damaged by fire. Most part of roof off.'
>
> **0036: Peers' Inner Court**. 'Back room on first floor and contents damaged by fire, heat and smoke and water, ceilings and contents under water.'
>
> **0036: Victoria Tower**. 'A quantity of builders' material damaged by fire, heat, smoke and water at front of building.'
>
> **0039: Little Dean's Yard**. 'Incendiary bombs. A building of one and two floors about 180 x 40 (used as assembly, lecture and classrooms) severely damaged by fire and roof off.'

0039: Kings College. 'A building of 2 floors about 180 x 30 (used as dormitories and recreation rooms) severely damaged by fire and roof off.'

0040: Ashburnham House.* 'Building of 3 floors about 60 x 40 (used as class rooms). Top floor and contents damaged by fire and most part of roof off, ceilings and contents under water.'

0041: Kings College Masters House. 'Building of 4 floors and basement about 50 x 30ft (used as residential quarters) about one half of top floor and contents damaged by fire and part of roof off, rest of floor and contents slightly damaged by fire, heat, smoke and water.'

By now Canon Don could see 'that our own nightfighters were busy in the bright moonlight, as the [AA] guns were often silent and machine-guns could be overheard'. Even so, he noted later in his diary that 'high explosives were coming down at intervals'. In her brick control room, Reenie Carter jumped at a 'big wallop' nearby. 'We were all thinking, "uh-oh, that was a bit close". They often came in threes, the bombs, so I wondered where the next one would land. It exploded closer still to us and as we waited for the next bomb we looked at one another as if to say, "Is it going to be us?". There was a great bang and all the windows along the top of our shelter came in and everything inside rattled and shook. It took us a few seconds before we realised we were all right and then we started asking each other where it had fallen. There was a direct line in the control room to the Houses of Parliament as we had some firemen stationed inside and they soon called through to say, "Well, that was us but we're all right".'

The bombs had killed Arthur Stead and Gordon Farrant, two auxiliary policemen stationed in the octagonal stone turret at the south-east corner of the Royal Gallery who moments earlier had requested help in tackling a fire on the roof. Mr Bloomfield, one of the Custodian's

* Ashburnham House, named after William Ashburnham, the Cofferer of King Charles II and belonging to Westminster School.

staff, had bounded up the steps to the turret trailing a hose which he had given to the policemen. Then he'd leapt down the steps, three at a time, to turn on the water hydrant. The bomb descended more quickly than Bloomfield and sliced off the stone turret, sending it and the two policemen crashing on to the courtyard below. Bloomfield helped rescuers sift through the mountain of rubble to retrieve the bodies of Stead and Farrant and, more importantly, free the blockage. Just one obstruction in the 5 miles of corridor in the Palace of Westminster might have dire repercussions for its 700 or so ARP staff. Two further bombs dropped in quick succession, both at the southern end of the Palace near the Victoria Tower. The first plummeted through the slate roof over the principal doorkeeper's room of the House of Lords at its customary 500mph, ripped through the floor and that of the Lord Chancellor's messengers' room immediately below, continued down into the ARP store room on the ground floor and smashed into the kitchen below. The bomb didn't explode, but the tremendous rush of air ripped the heavy iron door from its hinges and hurled it into the adjoining room where Captain Edward Elliott was comforting his wife, Joan, and his daughter. The 54-year-old resident superintendent took the full force of the deadly projectile, and died unaware of what had hit him.

Mr S. Hartopp, a member of the Lord Great Chamberlain's staff, followed the flight of the second bomb as if he were a batsman watching the ball leave the bowler's hand. He was chasing incendiaries in the Law Lords' Corridor, near the House of Lords' Chamber, when he saw the aircraft come in low over the Abbey and drop its bomb. Hartopp stood transfixed, a sandbag held above his head and a hissing incendiary at his feet, as he watched the path of the bomb. It exploded and an avalanche of debris roared to the floor below, taking Hartopp with it, alive but badly wounded.

All the while incendiaries were falling incessantly like April rain on the southern end of the Palace. Dozens of firewatchers scrambled along slanted roofs and edged along narrow ledges to extinguish them before they could do any damage, except on the Victoria Tower. This had been girdled by scaffolding since 1936, to the consternation of the Fire Brigade, who pointed out that the wooden walkways around the 330ft Tower were ideal timber for German incendiary bombs. But firemen's concerns were brushed aside; it was far more important that the renovation to the Tower's stonework should continue.

Now the scaffolding was on fire and the Tower was imperilled. Victor Goodman, chief ARP officer whose other job was Principal Clerk of the Judicial Office of the House of Lords, led a firefighting party inside to tackle the blaze from within. Outside police sergeant Alec Forbes was still groggy after being knocked down in the blast that had killed two of his police colleagues. Unaware of the efforts being made to quench the fire in the scaffolding, Forbes slung a sandbag over his body and started to scale the narrow ladders that zig-zagged 300ft to the top of the Tower. He put out one incendiary, then another, and kept climbing the ladders that linked the levels of scaffolding. As he neared the seat of the blaze, in the south-east corner, Forbes met Goodman's party already hard at work. Happy that they had things under control, Forbes shrugged and retraced his steps with the sandbag still perched on his shoulder.

Incendiary bombs now fell at the north of the Palace, near the House of Commons, and the west side of Westminster Hall was also on fire. ARP staff started to run north along the corridors of Westminster Palace to meet the new threat. Outside Canon Don pushed open the steel doors to the shelter in the College Garden. The lights had as usual been put out at 11 p.m. on the orders of Dean de Labillière and the only noise in the fuggy gloom was the gentle hum of the air conditioning. At the far end of the shelter, Jocelyn Perkins was lying on a bed writing by torchlight. Around him were a number of the Abbey's staff including the tall and wiry frame of de Labillière and Archdeacon Donaldson and their wives. Don crouched down next to de Labillière's bed on the left near the entrance and briefed him on the situation. 'He got up and joined in the fray,' recalled Don. 'The Deanery was by this time alight and the flames spread from [Canon Michael] Barry's house to the other houses in Little Cloisters.' But by 1.a.m. there was little anyone could do. 'The water supply gave out,' Don recorded, 'and there was no pressure in the hoses. Barry finally rang up 10 Downing Street in desperation saying that unless help came the Abbey would go up in flames.'

The message from Barry was relayed down the private line to Ditchley Park where it was taken by Churchill's duty secretary, Mary Shearburn, in her shoe box of an office near the front door. She typed out the message and handed it to Churchill who was sitting in front of a film. The reply was swift and unhelpful: 'The Abbey must be saved at all cost.'

Away from the Abbey, north, south, east and west, the streets of London were becoming clogged with the dead. Twenty-four in the Alexandra Hotel, Knightsbridge where two bombs brought five floors crashing down; a similar number at the Turner Buildings in Millbank on the Embankment, just up from Parliament, when a 1,000kg bomb punched through the apartment block and exploded in the clay foundations where once Millbank Prison had stood; in Hackney at least twenty died when an HE demolished flats Nos. 45 to 96 in the Columbia Buildings; nine were killed by a direct hit on the Halton Mansions in Islington; five people sheltering in a tram tunnel at Lancaster Place were blown to pieces when a massive high explosive bomb ploughed straight through the road and exploded in the tunnel; twenty-five died in Bournemouth Road, Peckham, their ages ranging from 8-month-old Marian Oliver to 79-year-old Emily Humphreys; eight were lost when a bomb dropped on the London County Council

Weights and Measures Office off the New Kent Road, including 47-year-old firewatcher Les Smith who had been awarded a Military Medal in another war a quarter of a century earlier; two 500kg bombs fell on the west side of King's Cross Station at 1 a.m., killing twelve, and wrecking the grill room and bar, many of the general offices, and the booking hall. As the bombs dropped, a signalman had stood in his cabin aware of a 'terrible rush of wind ... I stood over the levers and put my fingers in my ears'. When he came to he 'rang the controller and told him there was some trouble'. At exactly the same time a German aircraft dropped its 1,000kg bomb on Barking, the only one to fall on that part of London this night. It was either a gross error of judgement by the bomb aimer or an attempt to lighten their load and shake off a British nightfighter. The bomb landed in the back garden of 183 Keith Road. John and Selina Curtis and their three children, Alf, Elsie and John and their daughter-in-law Ethel, still enjoying a family knees-up, were all killed. Job Drain, VC, and the other people in the communal shelter a few yards away and into which the Curtis family normally went during a raid, were unharmed.

Gladys Jenner's normal routine during a raid was to go to the Peek Freans shelter underneath the railway arches in Keeton's Road. But it had been too cold tonight, too cold and too lonely without her dad who was in Buckinghamshire visiting her younger sisters, so she and her 5-year-old brother and her mother stayed put in their house in Clement's Road. At about 12.45 a.m. they changed their minds. Gathering up some blankets, the three of them left their house, walked along Clement's Road and turned right into Keeton's Road. They quickened their step. The uneven drone of the bomber's engine grew louder. They broke into a trot. The aircraft seemed right above them. They started running as fast as they could. They reached the gate to the shelter. Two security guards jumped in front of them. 'Stop there! Stop there!' they yelled. Gladys heard a 'strange whizzing sound, it wasn't a whistle'. The bomb dropped on top of the shelter. The lucky ones were those killed instantly, blown apart or their lungs shattered by the blast. The unlucky ones were those like Gladys's friend Jim Gill. He had returned on leave hoping to go to the pictures, but as all his friends were spending the night in the shelter he'd come along to catch up on all the gossip. 'We stayed in the gatehouse,' remembers Gladys. 'The security men all knew my dad so they made us tea and told us to wait with them but we could hear what was going on.' They heard the clangour of bells and they heard the shrieks of the dying. 'Amid all the screaming I remember hearing Jim's pleas, "Let me die," he was saying, "I know what's happened to me. Please let me die." Rescue workers dug through the rubble with their bare hands, looking for his legs.'

THE bombs were no more arbitrary than usual on 10 May 1941. They picked their victims at random, indifferent to sex or age or godliness. One of the few bombs to drop in Croydon fell on the bowling green at the back of Gonville Road, just as 29-year-old Joan Rogers opened the window to look at the red sky over London. She was killed by the blast. At 15 Mann Street, Southwark, a couple were sitting either side of their kitchen table when an explosion dislodged the chimney breast above them. It fell noiselessly shattering the table but leaving the pair without a scratch. In Walworth, 19-year-old Hereward Barling was guiding a doctor to an incident in Rodney Road when a bomb dropped, killing Barling instantly but sparing the doctor. In Clapham, 15-year-old Maggie Meggs got out of bed to tell her parents they had to leave the house. Something bad is going to happen she told them. Muttering at the whims of teenagers, Mr and Mrs Meggs took their daughter to the shelter. On their return they found a 2ft spear of glass embedded in Maggie's pillow.

Some Londoners put their faith in their God to get them through the bombing; others, nearly 50 per cent in 1941, turned to astrology. One of the most vigorous believers in astrology was Thomas Mawby Cole, a 47-year-old retired businessman from Harrogate in Yorkshire who, in April, had addressed an astrology conference. 'Something staggering would happen on 11 May 1941,' he proclaimed with solemn conviction. Mawby Cole was in London on the night of 10 May, waiting for something staggering to happen, when a high explosive bomb sailed through the roof of the house he was in at Regent's Park and blew him apart.

Most Londoners were happy just to let fate take its course. 'The Blitz didn't make me religious,' says Bobbie Tanner. 'If it's going to happen, then it's going to happen and there's nothing you can do about it.'

17

GUY Gibson and Richard James climbed up into their Beaufighter through the panel in the bottom of the fuselage. James strapped himself into his harness in the swivel seat under the Perspex dome halfway along the fuselage. As usual for operational flights he was chewing a stick of gum and wearing his lucky red neckerchief. Gibson moved forward along the narrow catwalk, stooping slightly as he pushed his way through the armour-plated doors into the pilot's compartment. His seat was in the centre, in front of the bullet-resistant windscreen that sloped back close to his face. Twice already this night they had taken off from West Malling; at 9 p.m. to briefly test the Air Interception and at 11.50 p.m. on a patrol curtailed by interference with the radio transmitter. Third time lucky, they hoped. At ten minutes past one their Beaufighter's two Hercules engines took them up once more. 'We headed south to the Sussex coast,' recalls James, who was warmed in his dome by a hot air duct on the starboard side near his seat.

Over the intercom came the voice of a controller from one of the new radar stations on the south coast called Ground Controlled Interception (GCI). Gibson pushed a button on a box on his left and acknowledged the voice at the other end. The controller sat at a desk in Kenley in front of a giant cathode ray tube on which was painted the coastline of 29 Squadron's sector. All aircraft that came within range of the station produced blips which marked their position on the painted coastline. Other men and women at the station plotted the course, speed and height of those aircraft. This information was passed to a controller who used it to vector the Beaufighters on to the tail of the German bombers. The Identification Friend or Foe transponder (IFF) fitted on RAF aircraft enabled the radar stations to distinguish friendly aircraft from hostile ones.

Gibson's call sign was 'Bad Hat 17'. 'Bad Hat 16' was Pilot Officer Alan Grout and his operator, Sergeant Stanton. At 11.45 their controller had vectored them on to a Junkers 88 coming up from the south. Stanton made a couple of slight corrections on his radar and Grout opened fire when he was 100ft astern and slightly to the left of the bomber. Shells ripped into the fuselage and tail plane and it went into a steep dive that finished in the English Channel.

Gibson was eager to emulate Grout's success but it was now nearly 1.30 a.m. The first throng of bombers was on its way home and the second sortie was about to arrive. Gibson was told by the controller to orbit. In the back of the Beaufighter, James had his head under the leather visor trying to pick up a blip on the cathode tubes. Thousands of feet below, London looked like a fire grate with its red hot coals fanned with an enormous bellows as the flames were expanded and contracted by air currents rushing to fill the vacuum left by gases rising from the conflagration.

Gibson recalled that this night 'the flak over London was really terrible, not enough of it and not even accurate. All we fighter boys used to take no notice of it'. James worried when his pilot was in such a dismissive mood. On an earlier patrol over Hull, Gibson had disobeyed instructions by flying into the city's Inner Artillery Zone (IAZ) which, says James, 'meant the AA guns could loose off at anything they saw'. James had looked down from his dome, chewing his gum a little faster, as 'lots of tracer came up at us and followed us along. I called him on the R/T and said, "Sir, they're firing at us". Gibson snorted and said, "Not a patch on the Ruhr". He didn't think much of our Anti-Aircraft defence.'

Suddenly the controller's voice crackled over the intercom. 'Hello 17. Bandits ahead.' 'He gave us the right vector,' says James, 'and then when we were within about four miles of the German I picked it up on my radar and took over the interception.' For the next few moments James's eyes were Gibson's. He fired out instructions calmly and clearly over the intercom to his pilot as he followed the blip on the cathode tube and brought them within striking distance. 'Coming in from the left'. 'Range 8,000ft'. 'Hold your course'. On such a moonlight night, Gibson saw it early. 'I preferred darker nights,' says James, 'because while we could get closer to the enemy on a bright moon, so they could see us more easily.'

Gibson adjusted his safety straps and then squinted through the ring sight (gun sight) and simultaneously twisted the safety catch anti-clockwise on the gun button. As his thumb hovered over the button Gibson lined up the spot in the centre of the ring sight on to a small sheet of glass just in front of him. Then he checked the range and speed of the target using the ring. Secure in his swivel seat, James braced himself for the manic hammering of the Beaufighter's four cannons. Gibson jabbed down his thumb against the gun button. Nothing. He throttled back to avoid overshooting and worked his way into another firing position. Again he pressed his thumb down, more urgently this time. Nothing. They had a jam. Now the rear gunner in the bomber had seen them. He fired a burst at the Beaufighter. Gibson broke off the attack. James watched as the tracer 'floated past us like a string of red hot sausages'.

Soon after James guided Gibson on to the tail of a second enemy aircraft. Again the same emotions: excitement, apprehension, expectation and anger as the cannons malfunctioned. 'It was terrible,' says James, 'a great disappointment. I remember thinking as we returned to base that we should have been the first in the squadron to have shot down two Germans in a night but instead we'd got nothing.'

WHILE the Spitfires of 74 Squadron and 29 Squadron's Beaufighters had taken off, the Defiants of 264 Squadron had been kept waiting at West Malling. 'The increasing rumble of heavy gunfire and the continuous thud of falling bombs told us that London was being subjected to a major blitz,' recalled Flying Officer Freddie Sutton. 'And still no one was ordered into the air. This was absurd.' Seething under his ancient greatcoat Squadron Leader Arthur 'Scruffy' Sanders succumbed to impatience and 'disappeared towards the control tower to burn up telephone wires and start up some action'. He strode across the grass, the 'whiskers of which he was so proud' jiving in time to his step, and demanded answers. He returned in a few minutes and gathered his aircrew around him. 'We shall be completely uncontrolled,' he told them. 'You can fly anywhere you like in the sector and at any height. This is developing into a very nasty raid and there may be anything up to 400 aircraft milling around the sky, so for God's sake keep your eyes skinned.'

Sanders with Sutton as his gunner began their patrol at 12.24 a.m., turning south towards Beachy Head at 10,000ft. In the back of the turret of his Defiant, Sutton went through his familiar routines, 'checking and adjusting my gun sight and making sure my [four] guns were ready for instantaneous action'. He had discarded his tie for a silk handkerchief and his helmet 'hung on a convenient projection in the turret'.

The skies were crawling with Germans and twice in quick succession, Sanders wrote later, 'enemy aircraft passed very close to us in the opposite direction'. He called up Sutton on the intercom. 'I think we're in the wrong place. Let's go to London and have a party.' 'Suits me, sir,' replied his gunner. As they flew north there was little chatter between the pair. 'Anything he had to say was always witty and to the point,' said Sutton. Occasionally Sanders might loose off a few choice words at the unseen enemy, 'Hun taunting', as he called it. 'He would lean from his cockpit and yell insults,' recalled Sutton, 'ordering me to do the same. "Come and fight, you lily livered bastards!" ... all rather futile but it was good fun.'

Between Beachy Head and London they saw no sign of the enemy, just dense smoke billowing up below them from the city. But it was the hiatus between the first and the second wave of German bombers. Sanders 'proceeded to patrol along the south border of the Inner Artillery Zone at 17,000ft'. Sutton enjoyed flying with Sanders. 'He was very steady – in fact after flying with some of the boys he made me feel that I was being trundled around in a bath chair.' As the pair scanned the night sky for Germans, their frustration grew. Beneath them their people were dying in their hundreds. 'At every splash [of exploding bombs] we thought of the fire fighters and the ARP workers all of whom were the subject of so much ridicule in the early days of the war,' he mulled. 'Poor devils, they were justifying their existence now.' Round and round they flew, 'feeling absolutely impotent ... and as we watched our rage mounted'.

Suddenly a Heinkel He111 flew across the Defiant at right angles heading towards London. 'I chased the aircraft,' said Sanders, 'and it immediately dived away from me in a wide left hand turn.' Sutton had seen the German too. 'Tally ho, sir,' he roared. 'After the bastard!' Sanders pursued the German, 'in an almost vertical dive with engine at full throttle'. At 8,000ft the Defiant caught up. 'We were tucked under his starboard wing about 20 yards out,' recalled Sutton. 'I opened fire, and missed completely! The Hun gunner was belting away and I could see his tracer going over my head and way beyond the tail.'

Both aircraft continued to dive until they were 'travelling at well over 400mph'. Sutton was struggling to stay conscious, let alone 'deal coolly with deflection* and aiming problems'. It comforted him to know his German adversary in the dorsal turret was battling the same G force. At 4,000ft the two aircraft pulled out of their dives.

Sutton 'gave him a good burst in the right place' with his four Browning machine guns. 'Straight down went the Heinkel with both engines ablaze.' The pair watched as the aircraft screamed towards the ground. Inside the five crew might have been trying frantically to reach one of the two escape hatches, but the centrifugal force would have been too great. There were no parachutes, just a 'big explosion below' as the aircraft slammed into the ground it had come to bomb.** There was also, lingering in the cold night air at 4,000ft, the familiar stench of burning Heinkel. It was the sweet and sickly smell of the light alloy used in the airframe. Some RAF pilots found the smell slightly nauseating, others found it satisfying.

Sanders and Sutton returned to West Malling at 2.24. 'We gulped down tea at dispersal and compared notes with the boys,' remembered Sutton. They took the reports from their armourers and mechanics and gave them a cursory glance. Then the pair unwound as they always did after a night patrol, with a game of snooker.

* Deflection shooting was the calculation of the angle at which to open fire during an attack when not directly in front of or behind the target.

** Sutton said later: 'It took a fortnight to find the wreckage, for he had landed neatly in the Thames in the heart of London.'

18

AMERICAN journalists Quentin Reynolds and Ed Beattie had ignored the first German bombs to fall close to the ground floor restaurant in the Savoy Hotel. So had Larry Rue and his chess partner, the delectable Claire Luce, hunched over their board in Titch's Bar. Only when the kings and queens and pawns began to rattle and shake did they realize the explosions were getting closer. Reynolds and Beattie went to take a closer look from the back of the Savoy on the north embankment of the Thames. Reynolds described what he saw in a cable he sent *Collier's* the next day: 'The angry buzzing of the planes increased and the bombs rained down; literally rained down. We hugged the side of the building. Now the fires appeared. A large bomb dropped across the river burst in a brilliant orange flame seemed to die and then spurted high into the air. There were warehouses over there. One by one they caught fire. The fire engines clanged sharply and then the air was filled with a great hissing as the water met the flames.'

Beattie and Reynolds went back inside the Savoy and climbed the stairs to the first-floor room used by journalists as a meeting place. Bob Post and silver-haired Jamie Macdonald of the *New York Times* were there, so too Ben Robertson, Larry Rue and Claire Luce. They all scented a good story. Word came through that the United Press Building in Fleet Street had been hit by incendiaries. 'I'll have to get to Fleet Street,' Beattie said. Reynolds looked at him in disbelief. 'You're crazy. It's a mile walk and stuff is dropping all over.' Beattie shrugged. 'I'm in charge of the office tonight. We have records there and I want to get them.' He asked if anyone had a tin hat they could lend him. But no one had. Few journalists bothered to carry one. Reynolds had owned one once, but then he'd dropped it and it cracked in three places. Claire Luce asked Beattie if she could come with him. 'Sure,' said Beattie. 'If you want to be a fool.' Reynolds wished them luck as they departed to 'face a mile walk in the midst of the worst Blitz London or any other city ever had'.

Beattie and Luce dashed out of the Savoy on to the Strand. As one of Beattie's colleagues wrote later, bombs were 'falling like grain from a seed drill'. They ran 100yds east to the Aldwych. Passing the northern end of Waterloo Bridge they saw that a wing of Somerset House was on fire. At King's College a few yards further on smoke and flames spouted from the windows. But it was the sight of St Clement Danes Church just before Fleet Street that made them rear up like startled horses. Even a hard-boiled American journalist and a ditzy Hollywood actress were shocked by the enormity of what lay before them. This was one of England's greatest churches; it was on this spot that Danish Vikings had married their English women in the ninth century. Christopher Wren had erected the church in 1680, with its oak pillars crowned with cherubs and its stalls carved with angels. Dr Johnson had worshipped under its lead roof and he was depicted on the stained glass windows along with Boswell, Goldsmith, Burke and the Cockney flower girls who for centuries had sold posies of violets outside the church. At 1.35 a.m. a German bomber had dropped a breadbasket of incendiaries on top of the church. Firewatchers counted sixty of them. When Beattie and Luce arrived flames were writhing from every window and every door. Outside, the Reverend William

Pennington-Bickford and his wife Louie wept. For thirty-one years he had been rector of St Clement Danes, instituting the Oranges and Lemons Easter service in 1920 when children were given one each of the fruits as the ten bells played the old nursery rhyme. With a terrible ululation, the bells broke free from the tower and crashed into the flames below.

BACK at the Savoy Reynolds had talked his way on to the roof for a better look. 'I'm from the Air Ministry Intelligence Department,' he told the firewatcher. The man eyed up the American amusedly. 'You're from room 554.' Reynolds recognized the firewatcher as one of the Savoy's barmen. 'He let me stay a few minutes.' Rue joined his compatriot and began writing the copy that would appear in Monday's edition of the *Chicago Tribune*.

> From a rooftop during the night I saw again the huge semicircle of fires. I heard overhead the almost constant whining of airplane engines – not individual planes but squadrons. I could see the reflections of the fires in the Thames, silhouetting the outlines of famous buildings still standing.
>
> I could hear the whistle or swish of bombs as they descended. I could feel the tremor of the earth as the roar of exploding bombs numbed the eardrums.
>
> Searchlights played about and sometimes caught enemy planes in a box of light as they were diving at targets and lending a crescendo to the multiple noises and sounds of an air attack. It was even possible to see enemy airplanes at times as they crossed the disc of the full moon. The bombers themselves seemed to swarm in like hawks.

Reynolds too was at work. 'London was in flames,' he wrote. 'Across the river a solid sheet of leaping maddened fire banked the river for nearly half a mile ... there were flames on every side. I looked toward Fleet Street and winced. Two large fires were reaching up into the night. I had a lot of friends working down there.'

At his desk on the fifth floor of the Reuters Building, Hughes was writing to his daughter in Australia when the first bomb fell on the *News Chronicle* offices in Fleet Street at 12.45 a.m. For the last forty-five minutes he'd been on overtime, Hughes explained, at four shillings an hour. So the raid was rather a good thing 'as I shall be here till the All Clear at any rate (dawn is at 4.30 a.m.) and as I expect I shall have to go out in the All Clear to cover "Air Raid Damage", I ought to earn at least a couple of quid today'. From what he could see from his window Hughes reckoned 'poor old London is well and truly for it tonight and I expect there will be another nice mess in the morning ... but everybody is going on with their work as though nothing were happening. The papers must come out – that is the tradition of Fleet St and you can take it from me that no matter what Goering does to London tonight the tradition will be observed. I was up in the library a short time ago and you could hear a perfect hail of incendiary bombs falling outside the street. What a life!'.

The incendiaries that landed on the *News Chronicle* in Bouverie Street also flamed up Serjeants Inn, a tranquil backwater of Georgian houses and cobbled streets just off Fleet Street. The caretaker had gone away for the weekend, with the keys hanging from a chain on his belt, and so all the houses, including the former residence of John Thaddeus Delane, editor of *The Times* between 1841 and 1877, burned long into the night. The heat from the inferno in the *News Chronicle* offices turned the south side of the adjoining *News of the World* white hot. The sanitary pipes melted and paint peeled off the walls, but the staff stayed at their posts, adamant the paper would be published in a few hours. But then the *News of the World* was attacked by flames from the direction of Serjeants Inn. First a printers and publishers in

No. 4 Bouverie Street caught fire, then the *News of the World* began to burn. Some of the paper's staff fought the fire with stirrup pumps allowing colleagues precious minutes to salvage paper and equipment from the building and move to the standby premises across the road at No. 8. Then the water ran out. Newspaperman Ernest Riley recalled that 'the fire roared up our area as in a blast furnace; the wind licked it into our building and we had to abandon it'. Journalists in the *Scotsman* on the corner of Bouverie Street and Fleet Street saw the flames getting closer and evacuated their typewriters and telewriters to Bouverie House.

John Hughes left his desk for 'a quick look-see out the front door'. As he climbed the stairs back to his office on the fifth floor he passed some of Reuters' firewatchers, who had just stamped out some incendiaries in the sixth-floor library. Hughes told his daughter that 'Ludgate Hill was on fire from both sides for about 100 yards from the Circus. New Bridge St was badly alight from the corner down towards Blackfriars, Shoe Lane was a mass of flames … St Bride St was on fire, St Andrew St from the junction of Shoe Lane to Holborn was on fire … generally there was such a schmozzle as never was'.

Among the buildings in Ludgate Hill hectored by the flames was Cassell & Co., the publishers, in La Belle Sauvage Yard, the same spot where for four hundred years a coaching inn had stood in which plays were staged and, for a time, drinkers had been able to take a ride on a rhinoceros for two shillings. Since early on in the raid the solitary firewatcher on duty had zig-zagged around the upper floor scotching out incipient fires. With each incendiary he doused, the firewatcher struck a blow for the hordes of British children looking forward to the next edition of *Chums*, the boys' paper, and *The Girl's Realm* and *Little Folks*, just two of the many annuals published by Cassell's. But then bad luck. With an ostentatious whoosh, an oil drum in the adjoining building was ignited by an incendiary. Soon that fire was reinforced by others from the direction of the Old Bailey. The firewatcher stuck gallantly to his post as flames began to encircle Cassell's. Only at the last moment did he flee for his life, escaping into Fleet Lane and leaving La Belle Sauvage and the ghosts of drinkers past to the inferno.

From Ludgate Hill the fires swept over Ludgate Circus, into Fleet Street, Shoe Lane and up the narrow Nevill's Court, climbing over the oldest domestic houses in London. No. 10 had been built in 1662 by a city merchant who had prospered under King Charles II. Other plaster-fronted houses had followed immediately after, seventeenth-century curiosities in the bosom of London, unmolested by the passage of time. The Great Fire in 1666 had been held at bay by the Court's front gardens. The fire of 10 May 1941 wasn't as compliant. After Nevill's Court the flames spread to Fetter Lane, threatening the Gothic-style Public Records Office and the *Daily Mirror* offices in Geraldine House, another monument to vanity recalled one of its journalists, 'that rose in tiers like the decks of a passenger liner, the printing machines in steerage the bosses in the staterooms above'. How the British Establishment would have squealed with glee if they knew the *Mirror* was in mortal danger. For months the *Mirror* and its Sunday sister, the *Pictorial*, had pilloried the country's elite for wining and dining while London's poor died under German bombs. Churchill was enraged, telling Cecil Harmsworth King, the papers' managing director, by letter that his paper should not 'try to discredit and hamper the Government in a period of extreme danger and difficulty. Nor ought it lead you to try to set class against class and generally "rock the boat" at such a time'.

King, so he told his diary, thought Churchill was being 'pretty petty' and rather flattered that their criticisms had 'got very deeply under his skin'. But he toned down his attacks when the Government threatened the papers with obligatory censorship.

Nothing, however, could quell King's innate disdain for the old class-ridden London. He told friends that the Luftwaffe 'by clearing away so many sordid buildings ... gives London the chance of a fresh start'.

Whatever one thought of the 40-year-old King, he was leonine in his courage. But there was no way the Government would permit his travelling overseas as a war correspondent. He channelled his energies and abilities into firewatching, and as the flames in Fetter Lane began to taunt the *Mirror*'s offices, he climbed on to the roof of Geraldine House. King spotted that some incendiaries had landed on the roof of the paper's old offices across the road in Bream's Buildings. 'I led a party over the roofs from the main building and we threw up sandbags to a man who appeared on the roof above us. He put the sand on the flames and put them out.' Inside the flames were restive and only one fireman was trying to calm them. 'We went down to the lower floor,' wrote King in his diary, 'forced open the lift door and directed a stream from a stirrup pump on to the flames; meanwhile the flames were breaking out again on the floor above.' Finally they subdued them, but elsewhere in Fetter Lane flames were coursing through several shops, some factories, a distillery and a printing works. The Royal Scottish Corporation was evacuated. King rejoined his men on top of Geraldine House, just after they had stamped out an incendiary on one of the tiers of wedding cake. 'The Germans overhead were in such large numbers that you could not distinguish the sound of individual planes,' wrote King, who later reckoned that every minute for five hours a 'bomb could be heard rushing or whistling through the air'.

On top of the *Mirror,* King and his firewatchers closed ranks, like a square of nineteenth-century British soldiers standing firm against a horde of screaming dervishes. Incendiaries and bombs attacked them from all sides. 'Once when I was on the roof, one dropped fairly near, and one could see the flash and the column of dust from the explosion. Fires by this time were all around us.'

They screamed for reinforcements as the breeze drove the flames nearer to the square. Guy Bartholomew, the 63-year-old editorial director of the *Mirror*, trotted into view wearing his AFS uniform and wiping the dust from his black, thick-rimmed spectacles without which he was blind. King despised him. Now he pleaded with him to call for fire engines. Bart promised him ten were on their way. One arrived a while later, 'not really for us but to protect the Public Record Office'. Fortunately they managed to eke enough water out of a pipe connected to the mains to fill a canvas dam outside the Public Records Office. They steeled themselves to repel the decisive attack but the breeze bridled its vigour, then veered off in search of a less defiant target. With Geraldine House out of danger, King stood on the roof looking out over London. 'Fires by this time were all round us – big ones and growing, especially a very bright one near the corner of Chancery Lane and Holborn.'

19

LEW White was sitting right on top of the fires that Cecil Harmsworth King could see. And from his position 'on the roof of Lincoln's Inn library', Lew could see the blaze under the feet of King. 'I remember vividly looking at the rest of London from the roof and thinking how light the city looked. It was literally lit up by the fires from incendiaries.'

Incendiaries had been spilling on to Lincoln's Inn since three minutes past midnight, when a shower deluged the offices of Chamberlain & Co, a firm of solicitors at 1 and 2 Stone Buildings. Lew and his colleagues sprinted the 50yds up the rectangular courtyard from their sub-station in No. 7 and contained the fire, while a request was made for a engine with a 100ft turntable ladder. Half an hour later, at 12.30, a stick of three high explosives fell on the Royal College of Surgeons in Portugal Street. There was a sub-station in the college but all its pumps were assisting with the fire a few hundred yards east in Lincoln's Inn. Reinforcement crews from south London arrived to tackle the blaze. Firemen dashed into the building and recoiled in horror. Scattered among the fallen masonry, covered in that perennial yellow dust, were skulls and thigh bones, even human hearts. It was several minutes before they realized these were college specimens freed by the blast. Some of the most valuable skeletons, including those of

Jonathan Wild, the criminal genius hanged in 1725, and the 7ft 10in Irish giant Charles Byrne, had been removed to the country, but other exhibits were consumed by the flames. The firemen were too late to save the mummified corpse of the deliciously wicked Mrs Van Butchell who, in her will of 1775, had stipulated that her husband could only take control of her estate if she remained above ground. So Mr Van Butchell had her embalmed and gave her an armchair in her favourite corner of the sitting room.

At 1.15 incendiary bombs fell on Old Square, Lincoln's Inn, on the offices of Finlay Campbell Kirkham barristers. They burned through the roof and set fire to the top floor, the flames casting eerie shadows on the Jacobean Gothic arches of the chapel in the south-eastern corner of Old Square. Bombs from the same canister also dropped on to the roof of the red brick neo-Gothic library, illuminating with their white-blue flames the immaculate lawns below. Inside the library were 70,000 law books and the ghosts of seven prime ministers, from Robert Walpole to Herbert Asquith. 'It was quite tricky to get up on to the roof,' says Lew. 'It wasn't that high but we had to get up over the battlements and the chimney stacks on to a flat ledge.' They manoeuvred the turntable ladder into position, avoiding the big bay windows, and 'ran up the ladder on to the roof carrying buckets of sand to put on the incendiaries'. Lew remembers that there was a discussion about whether to rig up a pulley to get the sand up on the roof or to continue doing it manually. 'But we didn't have enough men or enough time to do that. There was only one pump at the library because the rest were out fighting other fires. We just knew that we had to put those ruddy incendiaries out because once they take hold, that's it.'

Once, twice, three times, Lew descended the ladder and brought up more sand for the firemen clambering across the library roof dousing the incendiaries. 'Climbing up the ladder with only one hand was very, very frightening,' he says, 'but as I went up past the library I thought what a tragedy it would be if we lost all those ancient books.' With the fight to save the library won, Lew leaned back on the oyster-grey sloping roof and rested. Sweat ran down from under his steel helmet despite the cold. He looked south towards Fleet Street and the river. Behind him, on the northern side of High Holborn, another of London's four Inns of Court, Gray's Inn, was being annihilated.

Gray's Inn, with its well-ordered squares and its chambers with their own private gardens, had atrophied somewhat throughout the preceding months. Francis Bacon, a pupil at the Inn in 1579, had been hurled from his plinth in the South Square and the red-brick hall and the library had both been patched up. The barrage balloon that loitered above the Inn had been a feeble guard dog.

When the first incendiaries had dropped in the early hours of 11 May, they were snuffed out with alacrity by the Gray's Inn's fire-watchers who had been sleeping on mattresses in one of the class-rooms. As more bombs fell, the Inn's head porter, Mr Ivey, directed his firewatchers towards the static water tanks sunk into the Inn's terrace. A sergeant-major in the Coldstream Guards during the first war, Ivey was delighted to be back in action. With calm authority he led a charge towards the demented flames engulfing the east side of the South Square along Gray's Inn Road. They kept charging until the water in the static tanks ran dry. Then Ivey ordered his men to fill buckets from the hot water taps in the Robing Room, and he fortified them with mugs of cider drawn from a cask in the cellar under the Hall. But their courage was to no avail. By 3 a.m. the Chapel, together with its small glazed turret, had burned to the ground. The Hall followed, 347 years after Shakespeare's *Comedy of Errors* had been performed there, and, in the library, all that remained of 32,000 books was a towering pyre of ashes.

Lew White watched fascinated as Gray's Inn burned. 'I could see flames spurting up and it wasn't hard to guess where they were coming from,' he says. 'Gray's Inn had led a charmed life up to then but it got it this night. But I was more concerned about the high explosive bombs dropping all around me. Some of them were so close they sounded like an express train coming out of a tunnel. By now I'd learned to judge their distance and if they weren't going to hit me I didn't bother about them. But it kept me on my toes.'

Back at the offices of the *Daily Mirror,* director Cecil Harmsworth King watched in dismay as the 'Record Office water gave out. We scrounged around, found a small pipe in the yard which worked directly off the main and which still flowed … though very slowly'. The water mains in Fleet Street had virtually run dry. A bomb hit a gas main just outside the *News of the World* and a grumbling roar sent every manhole cover in Fleet Street rocketing into the air. Plumes of whiteish-yellow flames spurted 15ft into the air through the covers. A policeman sent a message to the Gas Company asking for a mobile repair squad, before pacing up and down Fleet Street reminding people not to light up a cigarette. Then he stopped for a cup of tea at the YMCA canteen van that, with a stroke of good fortune, had not been parked over one of the manholes.

A repair squad of gas engineers soon arrived – 'pluggies' to the firemen. But the fracture in this pipe, like most gas fractures caused by high explosives, was too complicated to plug until later. With the main isolated the engineers began 'bagging' the pipe at a safe distance from the irascible flames. Ignoring the bombs, the engineers drilled into the road either side of the fracture. More incendiaries spangled Fleet Street as the engineers cut holes in the main and worked in bladders of canvas-covered rubber. They inflated each bladder, as if pumping air into a football, and the pipes were sealed temporarily. They would return to plug the mains when the bombs had stopped falling. For now they had to dodge their way through the burning streets to bag another fracture.

Then a real catastrophe struck Fleet Street. 'Local correspondents were telephoning some "nice" stories,' Hughes told his daughter, 'e.g., direct hit on Wardens' Post at Putney, and Chief Warden killed, German bombers down at different places, when all our phones went dead.' A repair squad was despatched, but a break in the telephone line was far more tortuous to mend than a gas fracture. The engineer hunkered down in his manhole, like a soldier in his foxhole during an enemy barrage, only he had to fix the telephone cable by the light of his torch. In the lead sheath of each cable might be as many as 3,000 split wires. Until the engineer patiently located and reconnected each wire Hughes knew he and every other journalist in Fleet Street was 'practically marooned from news'. Minutes before the phones went dead a report had reached Hughes that Westminster Abbey was on fire. He badgered his night editor to let him go and chase the story. But the night editor, Mr Cole, told him to go and get some sleep and 'if anything big breaks I'll wake you'. Hughes stared agog at Mr Cole. 'By far the most ferocious and sustained fire Blitz on any world city in history had taken place and was still going on but Mr Cole said he'd wake me up "if anything big breaks!"'

In the Savoy, news of the telephone catastrophe was broken by Joe Evans, chief London correspondent of the *Herald Tribune*. Quentin Reynolds recalled that Evans 'came in and dropped wearily into a chair. "About all the telephone lines are out ... can't get the Ministry of Information [but] the nightfighters got fourteen before one o'clock".' When Evans had recovered his breath he told his colleagues about his narrow escape, less than a month after his predecessor, James Minifie, had been blinded by flying glass during a raid. Evans had been touring the streets in the bureau's car when 'a stick of bombs burst nearby'. The driver, James Nener, was taken to hospital with cuts to the face, and Evans had decided upon a tactical withdrawal to the Savoy.

At the *News of the World* Ernest Riley and his colleagues continued to put the finishing touches to that morning's papers, while around them 'we were ringed with a scorching Gehenna of conflagration. There was a line of fire from Fetter Lane to Shoe Lane in the north, and from Serjeants Inn to Temple Church (a blazing torch) in the west'.

THE Temple began at the Embankment, from where its lush gardens with the kneeling black figure supporting a sundial sloped north for a few hundred yards until they reached the brick-lined courtyards of the Inner and Middle Temple with their plane trees and flagstones. By 1941 the Inner and Middle Temple were long established as two of England's four Inns of Court, and the buildings were redolent of an illustrious heritage. The 100ft-long Middle Temple Hall, erected in the 1560s, boasted a wooden serving table carved from Sir Francis Drake's ship, the Golden Hind. Beyond it was Brick Court, home to a litany of notable residents including Oliver Goldsmith, the Irish writer and friend of Dr Johnson, and the novelist William Thackeray from 1853 to 1859.

The Temple's Fountain Court had been mentioned by Charles Dickens in his novel Martin Chuzzlewit, and the Inner Temple Hall, although rebuilt in 1870, retained its two Elizabethan doors and a 1709 painting of Pegasus leaping from the summit of Mount Helicon. The Middle Temple Gateway was the work of Christopher Wren in 1684. The Lamb Building, named in honour of Charles Lamb, was constructed in 1667, one year after the Great Fire of London which burnt itself out just yards away. Most venerable of all was the 750-year-old Temple Church with its pepper-pot roof and its effigies of nine thirteenth-century knights.

AT 1.50 a.m. on 11 May a flight of incendiaries speckled the church roof above the recumbent knights. For fifty seconds they burned at 2,500°C. A group of firewatchers arrived, among them J. W. Morris, a barrister at the Inner Temple. 'We ran there and began placing the red fire escape near to where the fire was on the roof,' he said later. A fireman appeared and told them not to bother, 'as there was no water available. He was not able to do anything. I went to my chambers and tried to ring up the Fire Service but I found the lines were all dead'.

Morris returned to the church, brimming with anguish. Couldn't the firemen do something to save the church? They told Morris that 'calls for help had been made but there were a great many fires in many places needing attention'.

Soon the flames had gallivanted across the south-eastern part of the church roof on to the Lamb Building. 'The best thing seemed to be to try to save Lamb Building and keep the fire localised on the roof of the church,' said Morris. He rallied his firewatchers and began searching the Inner Temple for water. They found a full tank at the top of a building. But in between trips with their buckets, more incendiaries descended. 'I cannot say at what stage Harcourt Buildings became on fire,' said Morris, 'nor at what stage the demolished Brick Court building was set ablaze.'

They thought only of their struggle to save Lamb Building. '[We] seemed to be succeeding but just as we thought this was so the fire broke out again near the ceiling when our water was coming to an end.' Morris ran downstairs to the court and urged a fireman to follow him back up. 'Using his axe he was able to get at this fire and with his help we managed to overcome it.'

With the fire beaten the firewatchers left the Lamb Building. Almost at once they felt an easterly breeze on their faces. Morris hadn't noticed a wind earlier in the night, but there was one now and it whipped the sparks from the church roof in all directions. For a few seconds the sparks climbed through the air before they slowed, as if looking for a landing strip, before falling towards the Temple with casual malevolence. Some landed in the garden, others on the felt roofs of neighbouring buildings. Morris and his men dashed pell-mell back and forth and up and down the Temple dousing the sparks before they took hold. 'From the various roofs we could see ever increasing numbers of fires,' he recalled, 'some of horrible proportions. There was a very big fire south of Fleet Street and east of the Temple.'

But the flames in the Temple were just as insatiable. Having devoured the roof, hungry tongues of fire now flickered inside the church in search of further prey. As Morris made his way down Inner Temple Lane he met the sub-treasurer, Roy Robinson, limping his way back from a first aid post, his lower leg stitched back together after shrapnel had sliced it open. 'In spite of his injury he could not be deterred from taking an active part again,' said Morris. 'He and I, and some others, went into the church with fire extinguishers and tried to check the fire spreading through the pews. It was really a forlorn hope. The heat inside the church was becoming intense, and all our efforts were quite unavailing.'

Another barrister, Stephen Benson, darted inside to salvage what he could from the 750-year-old church. He returned with a prayer book and an altar chair, and tiny globules of silvery lead from the roof that decorated the back of his jacket.

The breeze seemed to be in league with the fire for suddenly it renewed its brio, 'blowing streams and showers of golden-looking sparks on the patched-up roof of the Inner Temple Hall'. Morris begged some firemen to spray water on the fire. 'They did their best and got a hose pipe connected. At first there was a thin stream of water but it very quickly lapsed into a mere trickle'. The fires skipped on with contempt; first from the church to the Inner Temple Hall. Robinson and Morris gave chase and in one of the rooms off the hall 'pulled down one or two pictures from the walls. One, I remember, was the portrait of Mr Justice Avory: the frame was hot and burned our fingers as we pulled it'. Next the flames barged into the library, consuming half of the 90,000 books, then they overran the Inner Temple kitchens to the Cloisters and the south side of Pump Court, a cloistered little square where Henry Fielding and William Cowper, the poet, had chambers in the first half of the eighteenth century. In the centre of the court was a pump, next to an old fire cart. Neither offered any resistance to the inferno. Some of the flames from the hall punished Morris for his earlier gallantry by attacking the Lamb Building. 'It was terrible to see it quickly enveloped in roaring flames,' said Morris. 'It burned rapidly as though of matchwood.'

And all the time bombs continued to drop. A high explosive embedded itself deep in the Temple Gardens but didn't explode; incendiaries torched the Harcourt Buildings, Crown Office Row and the chambers in King's Bench Walk. 'There were moments,' reflected Morris, 'when it seemed as if there was no limit to the havoc being wrought ... it was a night of unremitting toil and the exertion of carrying buckets of water, pumps and extinguishers was great.'

BACK at the Savoy Hotel, with the telephone lines down and the correspondents' blood up, an argument was raging according to Quentin Reynolds 'as to which Blitz was worse, this one or the famous Wednesday one in April [16th 1941]'. As Titch served the drinks and suggested it was time a 'Death Ray' was invented that could zap the German planes from the sky, the American journalists finally 'agreed that this was the more severe [raid]'. Two more of their colleagues walked in to the bar reporting for work, George Lait and Red Mueller of the INS news agency. 'Their clothes were filthy and their faces heavy with soot,' wrote Reynolds the next day, describing how their digs had been hit by a stick of three bombs. 'They had pulled wounded from

Tonight, however, he was striding from pump to pump cajoling his men and firing off messages to HQ.

One of the messages sent from Peek Freans was a request for Bobbie's petrol lorry. Some of the pumps on the scene needed refilling. 'I was sitting in the cab of my lorry reading by torchlight when I got a call,' recalls Bobbie. 'It was amazing how much time I did actually spend just sitting around.'

She drove out of Dockhead Station in Wolseley Street and on to Jamaica Road where she headed east for 650 yards. There was a torrent of fire straight ahead from the Surrey Commercial Docks and the northern banks of the Thames had a lustrous quality that gave the Thames the colour of burnished copper. She didn't allow herself to think what would happen if any one of the incendiaries dropping all around her should skewer her lorry. 'I was carrying 150 gallons of petrol in two gallon tins,' says Bobbie. 'The tins were of the old-fashioned sort and they were stacked on wooden shelves on either side of the lorry. If I'd been hit it would have been all over pretty quickly.'

Bobbie drove through the entrance to Peek Freans and was instructed to replenish some of the trailer pumps. She jumped down from the cab and opened the rear of the lorry. Each two-gallon tin of petrol weighed 17lbs. She grabbed two and looked for the first trailer pump in need of a refill. Engines became rapidly warmed during a night and when officers weren't around it was common for firemen to light their cigarettes from the heat. 'The engines were always red hot,' remembers Bobbie, 'and all I had was a square funnel to pour in the petrol.' She unscrewed the petrol cap and inserted the funnel. Another flight of incendiaries swooped down close by. 'This was the trickiest moment, pouring in the petrol, because all it needed was a couple of drops on the engine and I would've gone up in flames.' The firemen backed off discreetly as Bobbie poured petrol into the chattering trailer pump. Not a drop was spilled at the pumps despite the quiver from their engines. 'They didn't take much petrol so it was soon full and then I moved on to the next one.'

Bobbie went from trailer pump to trailer pump, filling up the petrol tank and returning the empty tins to her lorry. 'I suppose it was very dangerous,' says Bobbie, 'but at the time you haven't got time to think about that otherwise your hands might start shaking and that wouldn't be good.'

The dead at the Peek Freans factory were laid out in a neat row and covered with blood-soaked blankets. They were the silent ones who troubled no one. It was the screams and moans of the wounded that upset the survivors. Gladys Jenner and her family remained in the gatehouse drinking tea and trembling gently but uncontrollably. 'Wardens came in from time to time saying who had copped it.' Gladys learned that Jim Gill had died after losing both his legs. After a time the screams became too much. 'We just shut the door of the gatehouse.'

The wounded were ferried to hospitals in the back of the auxiliary ambulances, which could be anything from an old converted delivery van to one of the more recent grey wooden boxes mounted on a Ford Chassis. Inside were four wooden bunks with grubby canvas curtains for rear doors. The auxiliary ambulances had no suspension but as ambulance drivers were forbidden from driving more than 16mph it wasn't a problem.

Most of the London Auxiliary Ambulance Service (LAAS) drivers were young female volunteers who laughed at their description in the press as 'Mercy Girls' or latter-day 'Florence Nightingales'. The women themselves, in a nod to the acronym of the service, liked to be known as the Lousy Amateur Ambulance Service. Many of them came from wealthy backgrounds and had no need of the weekly wage of £2 3s. Anne Spooner joined in December 1939 because the petite 24-year-old 'wanted to do something for the war'. She had the self-assurance of the privileged and the winsome beauty of the fortunate. 'For the first time in my life I met ordinary people because up until that point I had led a very sheltered life,' she says.

Anne had been born in India, where her father was a successful businessman, and she was educated in England and France. In 1936 she was presented at court to King Edward VIII as one of the season's debutantes. 'I didn't want to be presented but my mother did. I couldn't really see the point of it all.' So the slums of south London were something of a shock to Anne when she was sent there to do her training in early 1940. 'I was horrified at some of the poverty I found in the Elephant & Castle,' she recalls. 'We had some terrible cases to deal with there, young women trying to give themselves abortions.'

Now Anne was at the wheel of one of the ambulances waiting for the wounded to be brought out from the Peek Freans shelter. All around her bombs fell, some with a whistle, some with a scream, some with a rush, some silently on the end of a parachute. 'All we could do was wait for the wounded to be loaded into the back of the ambulance,' she says. 'We thought of anything but the bombs.' Anne's mind flitted between themes: the wonderful job the rescue teams were doing, the flute she was studying at Trinity College, tea and food; but the recurring thought was of her husband, Tony, an RAF pilot in coastal command.

The attendant helped load the four wounded into the back of the ambulance. She covered them each with a blanket and jumped in beside Anne. 'Then I just had to get them to Guy's Hospital as fast as I could.' The challenge for any ambulance driver during the Blitz was to get their casualties to hospital in the blackout guided only by a beam of light the size of a shilling through masked headlights. At least tonight as Anne drove steadily west at 16mph, on to Jamaica Road where a landmine had landed on top of Dr Fox's surgery at No. 157, she could see as clearly as if it were daylight. They arrived at Guy's Hospital a mile and a quarter later. 'We'd help unload the wounded and make sure we got our four blankets back because they were scarce and we needed them for the next people.'

The scale of the tragedy at Peek Freans produced heroes from unlikely quarters. Gladys remembers one of the company's delivery drivers, 'a drunk who spent most of his time in the pub after the evacuation of his wife and children' had been heading for the shelter a few paces in front of her. His gait was unsteady and the alcohol hung in the cold night air. Sometimes he broke into song, a drunken, defiant dirge that he directed skywards.

When the bomb exploded the drunk stood motionless for a few moments, surveying the carnage. 'He went and got his delivery van and started taking people to hospital,' says Gladys. 'The responsibility sobered him up and he did a magnificent job the whole night, making trip after trip to Guy's and St Olave's hospitals.' Once in hospital the wounded weren't yet immune from danger; St Thomas's had already been hit, so, too, the Children's Hospital in Westminster and St Luke's in Chelsea. A 50kg bomb had exploded in the west wing of St Bartholomew's in the City and another had landed just outside in Giltspur Street, in exactly the same spot as a bomb from a Zeppelin in 1915. St Olave's Hospital was hit by a dozen incendiaries early on; these were trodden out by Lieutenant Eccles, the officer of a local Home Guard unit who had won the Military Cross in the first war. Seconds after quashing one threat, Eccles was blown over by the blast from a high explosive. He rose groggily to his feet and, seeing there was now a fire in the hospital mortuary, went to deal with it carrying a stirrup pump. He was awarded an MBE for his courage.

YOUNG Tom Winter on the Redriff Estate, a mile east from Peek Freans on the other side of the Surrey Commercial Docks, recalls that 'it seemed to have become calm once more' just before 2 a.m. He waited expectantly for the unvarying wail of the All Clear siren. 'That wishful thought didn't last long as in the distance once again we heard the anti-aircraft gun barrage starting up again and the inevitable drone of the enemy aircraft as they headed our way.' The Monopoly set had been packed away now and there was nothing on the wireless. Tom's mum had stopped nattering with Mrs Nunn from the upstairs flat. Everyone was trying to get some sleep behind the blackout curtains. 'Suddenly all hell seemed to be cutting loose,' says Tom. A stick of HE bombs fell on the docks, watched on their descent by firewatchers in their raid observation towers 60ft high. Tom heard the 'thump, thump, thump' as they exploded. At the Cherry Garden Pier, a few hundred yards west of the Redriff Estate, HMS *Tower*, the base ship of the Royal Navy Auxiliary Patrol whose job it was to defend the Thames, was holed by a high explosive bomb. The dead and wounded littered her decks as she sunk into the glutinous mud of the river. Less than a mile east the 5,600-ton SS *Niceto De Larrinaga* was hit as she lay in Bellamy's Wharf laden with cereal. The fires that broke out in some of the holds took her crew and firemen until next day to quell.

Inside No. 136, Tom and his family listened to the sound of an aircraft engine right above them. They knew from the drone whose side it was on. No one said a word, as if they hoped that by keeping quiet the bomber would go away. 'We remained silent and prayed,' says Tom, 'selfishly it would seem, for it to pass us by.'

They heard the bombs come down, pushing the air before them. Tom counted six, each one 'absolutely terrifying us all'. 'The noise quickly developed from the whistling down sound to a rushing ugly noise like an express train about to hit you.' Everyone in the flat threw themselves under the large wooden table in the centre of the room. 'We arrived in an untidy heap under the table at the same time as six bombs crashed into the Redriff Estate,' recalls Tom. 'Not with individual explosions but seemingly one terrific, almighty and terrible ear-piercing bang!' Their noise of the blast was so great, the pressure so powerful, that Tom thought his skull would burst like a balloon. There were three solid concrete floors above the Winters but still the floor of their flat leapt a foot into the air. All of them cowering under the table 'felt the sickening blast and severe concussion of the terrific force'. For a few moments there was silence. Everyone, says Tom, was 'saying a silent prayer thanking God once again for our deliverance'. The first one to speak was Tom's mum, her composure soothing the fears of her children. 'Thank God we're all OK. I'll make us all a nice cup of tea. It will do us good.' She asked Tom and his brothers 'to pop out on the street and try to find out from someone what had happened and if we could help in any way'. Tom knew she was thinking of their father.

'As we looked out of the front door we automatically looked to our left because we knew from which direction the sound came when the bombs fell.' They were looking west, towards the Dock and the Humphreys' flat. 'We could see only clouds of dust and smoke and there was a reddish orange-coloured glow from something burning.' People were running past the door headlong into the billowing smoke. Tom could hear strange cries coming from the smoke. A man sprinted past and screamed at the children to stay indoors. 'We came in and told mum we couldn't see dad.' Tom's mum smiled weakly and poured out the tea. 'About 20 minutes later dad arrived,' recalls Tom. 'His face and hands were dirty having been in among the smoke and debris.' But it was his eyes, empty and uncomprehending, that frightened his children. There was none of their usual merriment. 'This was dad who up to now had always instilled confidence and calm in others,' says Tom. 'He explained that he could not stay too long as something terrible seemed to have happened at No. 53, the Humphreys' flat.'

His wife thrust a cup of tea into her husband's hands, as if to say 'not in front of the children'. Harry slumped into a chair and gulped down the tea, then 'he told us roughly what had happened outside a short time ago and where all the bombs had fallen'. There was extensive damage to the flats in Elgar Street near the shops but no report of serious casualties. Tom's dad looked at his son and told him his friend Ken was OK but there had been a 'serious problem' at No. 53. A bomb had landed at the back of the Humphreys' block and bored into the ground underneath the flats without exploding. Tom's dad and a couple of his fellow wardens had been the first to reach the flat. They brought out Ken, his mum and sister, alive but stupefied, and led them away to a first aid post. As they did they passed four rescue workers on their way to bring out Granny Humphreys, still trapped inside No. 53. The men all lived on the estate: 20-year-old Tommy Bull at No. 84, Frank Cole at No. 36, Charlie Simpson at No. 50 and 53-year-old Charlie Wynn at No. 238.

Like everyone on the Redriff Estate, the four all knew Granny Humphreys well. Their torchlights found her sitting in her favourite armchair on the other side of the front room. The floor between them and the old woman was fissured like a drought-ravaged field. The men told Granny Humphreys that her family were safe and well and that they'd have her out in a few moments. Tom's dad was on his way back to the flat when he heard a 'sudden rushing noise of lots of earth and furniture falling from inside the flat'. There was no explosion, no bang, just the sound of falling. Tom's dad ran to the door of the flat and peered in. 'Where the floor had been there was now a deep chasm and an eerie silence.' The ground had swallowed up Granny Humphreys and the four men.

21

AFTER being turned off his tram in Walworth Road at midnight, Joe Richardson had made his way towards his cousin's house in Hillingdon Street. He walked with a cool swagger at first, thrusting his shoulders in and out as if he were walking past a group of girls at the Trocadero. Incendiaries plopped and hissed and sizzled and pursued him down the road. 'I thought to myself "I've had enough of this" and I started to trot down the road.' Joe turned right and made his way to Hillingdon Street, just behind the Walworth Road. 'My cousins lived there and they had a decent shelter made out of concrete which I thought would be safe.' Joe banged on the front door but there was no answer. 'I went into the backyard and looked around but there was no one there so I went back on to the street and ducked into one of the street shelters in John Ruskin Street.'

For two hours he sat alone in the darkness, listening. Shrapnel from the AA guns on Camberwell Green pitter-pattered occasionally on the shelter roof and explosions erupted all around him. The mighty roar he heard at 1.11 a.m. was the bomb dropping on top of St Mary's Church in Lorrimore Square, half a mile north. The crack to the south was an HE landing in Wyndham Road. More distant rumbles were explosions to the east in the Camberwell Road and Wells Way and Picton Street. Although Joe didn't know it, he'd just heard the death knell of the Palace Café in Denmark Hill and Pearman's Garage in the Peckham Road. Then at seven minutes past two, Joe heard 'an almighty whooshing noise followed by a huge bang!'. 'I really thought that was my lot,' says Joe. A bomb had landed on a row of houses in John Ruskin Street, punching through the roof and catapulting a girder through the air. 'The girder had practically demolished this house,' says Joe. 'I got back inside the shelter and hoped they wouldn't drop any more near me.'

Half a mile north of Joe, the Veazeys had walked the short distance from the smouldering ruins of St Mary's up Kennington Park Road to the Elephant & Castle. It seemed to Joan that a night had never been so godless. 'The Spurgeon's Tabernacle was burning furiously.' She and her husband, Christopher, joined a small knot of people trying to put out the fires in a row of houses behind the Tabernacle. 'Between us we managed to get these under control,' she wrote in her diary. 'I told one woman I was expecting a baby and she stared at me and said "Good God".' Joan shrugged. 'Sometimes I wonder if my baby will survive or be born holding a stirrup pump.'

An hour after requesting 'pumps 50' for the Elephant & Castle, Divisional Officer Geoffrey Blackstone had nearly 300 firemen fighting the towering fires. Water was still being pumped from the four hoses in the Surrey Music Hall and the Thames at London Bridge. But with every fire that was brought under control another one suddenly rose up menacingly, flames expanding outwards; at the Northwood Spare Motor Car & Accessories dealers in Newington Causeway; at the Home & Colonial Store in Newington Butts; at the Zetter Football Pools offices in Newington Causeway; at the Esso Motor Spirit service station in Newington Butts; at the Radio Rentals store; at the Isaac Walton department store; at the Inland Revenue Office; at the French Polisher's; at Tyler's House Furniture store; at the railway bridge of the Southern Railway at Newington Causeway.

In the Walworth Road, the flat of Austin Osman Spare was one of an incalculable number hit. Hitler, once incensed by the refusal of Spare to paint his portrait, was taking his revenge as incendiaries set alight the painter's studio. As he gathered up his canvases, high explosives spilled from the sky overhead and blew his flat apart. Spare, whose work had once been compared to Dante's, was experiencing his own private inferno.

All along the Walworth Road to the Elephant lamp-posts buckled in the heat, their heads drooping to look like flowers in need of water. Paint on the fire pumps blistered and bubbled, forming a ghoulish red liquid with the water that sprang from overflowing valves. Firemen hunched low over the hoses, faces near the branch to gulp down precious lungfuls of air. The water from the hoses turned to steam almost the moment it hit the fire; its back spray splashed off the firemen's helmets and ran down the gas capes that were used as neck protectors.

Reg Richardson had been dealing with incendiaries on top of the Elephant & Castle pub when he was told to grab a hose and follow his district officer. This was what London firemen called 'getting at it', dealing with the fire at its red roots. Provincial fire brigades preferred to fight the fire from outside because of the risk of falling masonry. Richardson scrambled up the stairs of an office block. 'I went into the room with the branch and did the very thing we had been told at training school not to do. I pointed the branch at the ceiling and looked up to see what happened. What happened was that loads of dirt and hot plaster came down on to my face and into my eyes.'

Richardson was led out on to the street where he was handed over to some firewomen. 'There were so many firemen blinded by the sparks and bit of debris,' says Emily Macfarlane. 'We couldn't do much more than daub their eyes with what little water we had in a bowl but that was soon dirty and black.' One fireman dropped to his knees in exhaustion.

'I ran over to him and gave him some food and he looked up at me. He looked at his wit's end. All I could see were the whites of his eyes, his face was completely black, just like he'd been down the mines.' Emily overheard another fireman laughing maniacally while telling his pal about the woman he'd seen imprinted in the side of a building in Dante Road. The bomb blast had driven her into the wall with such force that her body had to be prised out. Where possible some firemen lashed their hoses to try and give their heavy arms and thighs a momentary breather from the pressure of gripping the hose. But it was a dangerous practice. A hose that broke its mooring would whip free with a terrifying swish. Any fireman caught by the brass nozzle of the flailing hose would go down as if shot by a rifle bullet.

At 2.15 a stick of bombs blew apart five three-storey buildings on London Road, along which ran the four hoses from the Surrey Music Hall. Lumps of stone were tossed across the road but none blocked the vital supply lines. Firemen exchanged relieved looks. Seven minutes later a high explosive sailed through the ruined roof of the music hall and exploded among the firemen still pumping out water from the reservoir. Seventeen men, full of the sap of the living, were killed instantly, their bodies hurled in all directions to be left lying in grotesque, twisted shapes. The engine of one of the three pumps was blasted across St George's Circus and through the wall of the Salvation Army hostel in Blackfriars Road, where George Orwell had dossed during the 1930s. The hostel was famous for its free Sunday breakfasts and dozens of down and outs were camped inside waiting for dawn. All that was known of the two vagrants killed by the engine were their names. What remained of Thomas Palmer was reckoned by officials to be 'about 60 years old'.

Blackstone came up from the Elephant to see the damage for himself. The seventeen firemen would be mourned later, but his immediate problem was the wreckage blocking access to the reservoir in the basement. Without his main source of water, Blackstone returned to report to Lieutenant Commander J. W. Fordham, in charge of Southern Division, who had recently arrived to take charge at the Elephant & Castle. Now they were down to just the one water relay from the Thames at London Bridge. Fordham called up fifty more pumps. Then he ordered more relays to be laid from the Thames at Westminster and Waterloo bridges using a new kind of emergency steel piping consisting of 6in diameter 20ft lengths that could be coupled together and were far sturdier than conventional hosing. 'The biggest problem at the Elephant that night,' says Emily, 'was rubble falling on top of hoses and ripping them open or blocking the water.'

At the Elephant & Castle Underground station the yardmaster emerged from underground to see if it was safe for people to return home. He stared, dumbfounded by the pandemonium before him. It seemed that a tidal wave of red-gold flame was bearing down on his station. He mustered five members of staff and asked for volunteers among the scores of shelterers on his station's platform. Two old men stepped forward. Disgusted, the yardmaster ran to the booking hall and fixed up the main hydrants so they could fight the agitated fire from the roof. 'I called out "Turn the cocks on" but to our dismay not a trickle of water came.' The yardmaster collected his thoughts as his men looked to him for a solution. The tap in the booking hall was fed by a large storage tank at the top of the building. 'Six of us made a chain gang up the stairs and through a window on to the flat roof, throwing hundreds of pails of water on to the premises all around the station. The heat was terrific and panes of glass were falling out of the windows overhead from premises over the station.'

All around them they could hear the din of falling bombs. Ones that dropped too close shook the building, and the men on the stairs grabbed the handrail to steady themselves. After half an hour the dull ache in their forearms and shoulders had turned to a sharp searing pain with the weight of every full bucket. 'Anyway,' recalled the yard-master, 'we stuck it.'

Suddenly they became aware of a different light coming through the windows at the back of the station. The yardmaster went to look and discovered 'a restaurant had caught ablaze ... we began to think our work was all in vain but still we stuck it'. Then fire broke out in the airshaft that ventilated the sub-station. The two old men grabbed stirrup pumps and put out the flames. Minutes later it was the third floor alight. 'We dashed up the stairs and broke down a door on the landing,' remembered the yardmaster, 'and tackled this with stirrup pumps.' Some firemen arrived to help but saw the railway staff had things under control and left.

At 2.30 a.m. flames were seen on the fourth floor, but again stirrup pumps won that brief battle. A tall driver hared outside to douse an incendiary that had fallen in the information booth and returned with his scarf on fire. A stick of HE bombs was seen to tumble from the belly of a German aircraft overhead and the railwaymen 'scrambled through the landing window and down the stairs in case they were meant for us'. They fell wide and the men laughed nervously and talked of close shaves 'but not ones with our names on'.

THE Elephant & Castle was luckier than a lot of Underground stations on 10 May. The first hit was Rotherhithe, set on fire by incendiaries at 11.33 p.m. At 12.10 a bomb exploded in the tunnel between St James's Park and Victoria. At 12.42 a bomb blasted a milk train off the tracks between East Acton and Wood Lane. At 1 a.m. the Circle Line was damaged at King's Cross by a 1,000kg bomb that ripped apart the concourse. At 1.35 platforms 2 and 3 at Aldgate were wrecked by an explosion. At 1.46 there was similar damage to platforms 13 and 14 at Paddington. Hit, too, during the night were Tufnell Park,

Hammersmith, Great Portland Street, Drayton Park, Moorgate, Clapham South, Victoria, Stepney Green, Surrey Docks and High Street Kensington. Outside Euston a high explosive bomb dived through the road and struck a tube train on the Metropolitan line.

But Baker Street caught it worse than most. The station was deluged with incendiaries early on that caused a fire on platforms 2 and 3. Then a stationary train on platform 4 went up in flames. People filled up buckets from the tap in the staff lavatory and extinguished the blaze. A train guard remembered standing in the main entrance hall 'looking towards a fire that was burning on the top storeys of Bickenhall Mansions when I heard a whistle getting nearer and one of the chaps shouted "Look out, bomb coming down".'

Some of the guard's colleagues dived behind ticket machines 'but I was slow as I was looking for a place to shelter when the bang came. My ears seemed to be bursting and it felt like a hot wind going by my face. Then something seemed to be tugging at me. I dropped flat on the floor. A glass window fell with a crash not very far from me. The lights went out, then up again; dust came down in clouds as I lay there ... I got up and had a look round. The others behind the machines were getting up as well and looking about them. One of the fellows said "that was a big one".'

THE fires from the Bickenhall Mansions were also visible to 49-year-old Vera Reid from her flat in Gloucester Place. Before the war the Bohemian writer had lived overseas in a radical community. She returned to London in the late thirties, wrote a book called *Gods in the Making* and worked for the WVS on their refugee committee. She also kept a diary 'to purge myself of the sight of so much pain and distress'.

Vera's entry for 10 May had started well. She'd been to a divine wedding in the afternoon – everyone had agreed that Barbara's dress was exquisite – and she retired to bed in the evening, kissing goodnight the elderly Dutch Jewish couple with whom she shared the flat. When the raid began she'd tried to sleep through it, 'but it got too bad'. She got up, dressed, and sat on a chair in the doorway with her head in her hands fighting the 'paralysing physical fear' she felt. 'I couldn't go among other people and let them catch it too,' she told her diary. Gradually she reined in the terror and joined her flatmates next door. 'Neither of them looked in the least perturbed. I felt ashamed and I suppose I looked all right for they made no comment. Then it began all over again. It's like a sinister vibration which shatters you to pieces inside. Was quite incapable of movement though I wanted to get away before the others were affected like a leper.'

Vera could hear the thuds of high explosives and the plop of incendiaries. She heard too the hollers of wardens and the ring of fire engines' bells. Then from the street below she caught the sound of a man laughing. 'A good hearty laugh with a touch of excitement and triumph in it,' she wrote. 'My congealed blood began to flow again. My heart melted with love for him, for everyone. Even for those chaps up in the brilliant sky. Perhaps they too had been frightened and would now also find release from the fear. That laugh was like a signal for soon afterwards a fireman knocked on the door and told us we had better get out as they thought the house would catch fire any minute and they could do nothing about it.'

Two pairs of hands led her to safety as 'clouds of smoke rolled across Gloucester Place in both directions … a gas main had burst and was sending a burst of light up to heaven … Bickenhall Mansions were on fire from end to end and as we scuttled across Marylebone Road part of the roof fell in and a cascade of sparks and pieces of flaming material shot up like one of those set pieces we used to see at the Crystal Palace when we were children. The great full moon swung above against the blood red sky'.

A mile and a half west of Vera in Maida Vale high explosives had supplanted incendiaries as the chief worry of Olive Jones. A couple of hours earlier she had performed heroics in her neighbours' garden. Now her house was 'rocked and shaken by the blast from HEs several times and one especially violent explosion sent the glass of several windows flying and blew in the back door'. While her maids whimpered in the corner Olive picked up the phone and rang the Relionus car hire company. Could a taxi pick her up and take her to the WVS offices in central London? 'They promised to let me have a car within an hour or so though of course "I would quite understand that in the circumstances there might be a certain amount of unpredictable delay".' Just round the corner from Olive a bomb landed on the BBC premises in Delaware Road, killing one employee and severely damaging the basement studios. Another tore through the roof of a block of residential flats in Clifton Gardens. Incendiaries charged through the gaping hole and soon the four-storey building was in flames. The crew of the first pump on the scene were inside when the top floor gave way and sent down a red hot avalanche. Three firemen were killed and two others badly wounded.

In Wormwood Scrubs Park, Bombardier Bill Church could see the German aircraft dropping bombs on Maida Vale. It was 3 a.m. and his 3.7in AA gun had fired scores of rounds at aircraft whose range was detected by their crude Mark I radars. Behind them were piled thirty wooden boxes, each one containing two more 28lb shells. The ground all around was knee deep in empty shell cases, smoke still rising from those just fired. All the time three bombardiers kept up a steady flow of fresh ammunition to replace the shells being fired. 'We'd fired so many rounds that the paint had burned off the barrels,' recalls Bill. 'But at least the gun's hydraulic oil hadn't boiled out which sometimes happened.' The loader shoved the next shell in the breech with his leather mittens. It was punched up automatically and then fired. 'You wouldn't know if you'd brought down a plane,' says Bill, 'because there were so many people firing you didn't know who'd hit what.' Sometimes they fired and 'a Very light would burst which was an identification from one of our nightfighters'. The gunners muttered darkly about the Brylcreem boys thousands of feet above them. 'The RAF were a bloody nuisance at times,' reflects Bill. 'A target came into view and then one of our fighters would come buzzing after it.'

Bill and the nine-man crew worked like automatons hour after hour. Loading and ranging and firing and loading and ranging and firing.* After a while, recalls Bill, the ringing in their heads was replaced by a sharp pain with every roar of the gun 'like someone pushing a knitting needle deep into your ear'. The smell of cordite was everywhere, in their hair, in their skin, in their uniforms. It mingled with their sweat.

Ten miles south-east of Bill Church's gun, in Sydenham, Reg Harpur was still watching the raid from his front garden. 'Judging by the [AA] guns the sky is full of Jerries now,' he wrote in his diary. A couple of air raid wardens walked past at 2 a.m. and Reg asked them where they thought the fire was that was causing an attractive red glow to the north-west. 'One of them guessed the fire as being London Bridge but I thought it more to the west.' A large piece of AA shrapnel screamed through the air and Reg scurried back inside. 'I read several chapters of the story of the Norfolk Farm and tried to imagine myself with the author, Henry Williamson, who describes the efforts to run a farm on the experience of an author and writer ... I went to sleep about 3 a.m.'

* On the night of 10 May London's AA Command fired a total of 7,000 rounds between 10.50 p.m. and 5.15 a.m.

22

AT exactly 3 a.m. in Westminster two high explosives fell in the garden of Eaton Square. On the north side of the garden were three hundred vegetable allotments. On the south side in front of flat nos. 40 to 48 was a trench shelter full of people. Leonard Eaton-Smith, Westminster's mayor and the man Canon Alan Don had proposed as the People's Warden of St Margaret's Church, had just entered the shelter on a morale-boosting visit, when the two bombs exploded. The 67-year-old mayor died where he stood, as did several others, though it was never established exactly how many.

Walter Elliot, MP for Kelvingrove, recalled that it was 'about 3 a.m as far as I can remember' when he finished putting out incendiaries outside his house in Lord North Street. He asked a policeman about the situation a few hundred yards north in Parliament Square. 'He said "The House of Commons is on fire". As a good House of Commons man it seemed to me that this was where I came in. I went down to the Embankment to walk along to the House.'

As Elliot hurried north he noticed how the Thames was striped orange and black with the fires from Lambeth Palace and St Thomas's Hospital on the south side. Ahead of him in the distance fires were 'flaring all over the City of London'. At the Palace of Westminster he found chaos. The unexploded bomb that had ripped through the ARP control room and crushed Edward Elliott meant a new headquarters had to be established. But this had to be done in the dark for the UXB had wrecked the Palace's lighting system.

As Civil Defence personnel felt their way through the Peers' Lobby and the Central Lobby, guided only by the spears of light from their torches, towards Big Ben and the southern end, incendiaries fell on the House of Commons and the east side of Westminster Hall.

Elliot stood under the Victoria Tower, still smouldering from the earlier fire on the scaffolding, and looked across Old Palace Yard to Westminster Hall and the carnival of flames on the roof where, for twenty-five years, Oliver Cromwell's skull had been impaled on a spike. He could see hosepipes slithering up the steps to the Hall at the St Stephen's entrance. 'I went to the top of the Hall,' Elliot said, 'and saw that the great timber roof was well alight.'

Chief ARP Officer Victor Goodman was aware of the situation but had other problems. On his way back through the Palace, having seen the flames gobbling up the roof of Westminster Hall and the House of Commons, Goodman's eye was drawn to a neat hole in the east side of the roof over the Chamber of the House of Lords through which the night sky was visible. Finding a similar size hole in the carpet underneath the roof, Goodman flashed his torch into the abyss and its butterfly beam picked out four glinting fins. He hurried back to the control room, where the phone was one of the few things still working, and requested immediate assistance from Civil Defence Control at Lambeth.

It was the call Chief Superintendent Charlie McDuell, commanding officer in London's West End, had been waiting for. He yelled to Leading Fireman Dave 'Dusty' Millar, a Scot from Inverness, to get the engines running on his staff car then turned to the man next to him, 26-year-old Robin Duff, a slim dark-haired reporter for the BBC.

'You want to come?' he asked. Duff's principal job at the BBC was using his precise clipped tones on the 'Imperial Echoes' programme. Now he was promised real excitement. 'In less than a minute the big swing doors were open and the appliances were on their way,' he wrote later. 'Not only from our station but from every post in the district ... we wasted no time on the road and there was no slowing at crossroads; our bells were going and it was up to other cars – if there were any about – to keep clear. More than once our driver swerved suddenly to avoid a new bomb crater, and glass and debris crunched under the wheels.' They roared past Buckingham Palace, 'strangely gaunt against a sky of flickering red' and into Parliament Square a mile further on. Duff recalled that as McDuell set up his HQ in New Palace Yard and began to marshal his men, he surveyed the scene. 'The fire was in Westminster Hall. The small rooms along one side that faces Oliver Cromwell's statue were well alight. There was another fire on the roof of the Victoria Tower but already firemen had that under control.'

As the dozens of fire pumps arrived at the Palace, McDuell ordered Millar into the House of Commons to make a situation report and listened to the briefing by Walter Elliot on the blaze in Westminster Hall. 'Vigorous attempts were being made to subdue the fire,' said Elliot, 'but it was moving further and further away from the platform at the south end of the Hall and it was evident that the whole roof was in extreme danger.'

The heat was now so great that the lead guttering on the east side of the Hall melted and the flames were given added impetus by the varnish used to treat the beams for deathwatch beetle. As McDuell listened to Elliot's report he heard a familiar refrain, a whistle, then a dull crack. There was a flash of light in the belfry of Big Ben and giant shards of glass from the broken clock face on the south dial fell like lances on to the ground hundreds of feet below.

Duff had missed the injury to Big Ben. Instead he got inside Westminster Hall 'through a small side door' and was trying to make out the extent of the damage so far. 'The building was filled with smoke and flames were licking frames of the windows,' he recalled. Duff coughed and spluttered and remembered for a moment the last time he had been in the Hall, to view the coffin of King George V as it lay in state in 1936. 'Now it was just another a building on fire,' he thought. As Duff emerged from the Hall into New Palace Yard he heard a bomb whistling from an aircraft with 'a roar that tells you this one hasn't got your name far off it. There was a crash and we found ourselves thrown along the road, the rubble falling on our backs. I looked up at Big Ben and saw only a cloud of black smoke. Then the chimes rang out'.

As Duff, Elliot and McDuell readjusted their attire they turned to look at the House of Commons, from where the explosion had come. 'Flames shot up … from the whole length of the chamber of the House of Commons,' said Duff.

Moments before the bomb had dropped Dave Millar was standing in the Strangers' Gallery overlooking the Commons Chamber. He ran his torch along the rows of green leather-backed benches and over the 13½ft Speaker's Chair, sinister in the shadowy light. Everything appeared in order. He closed the door and the world convulsed. The bomb somersaulted Millar down the passageway, his body thudding against the floor and walls. For a few moments he lay motionless. His head felt as if it were floating but someone was ringing a bell against the inside of his skull. He began to cough and spit out dust. Millar crawled away from the flames he could feel under him, nuggets of rubble digging into his hands and knees. By the time he had been helped back to the HQ in New Palace Yard, Millar had unscrambled his senses but he was unable to answer McDuell's question: he didn't know what had exploded in the Commons Chamber. It might have been an oil bomb, but they didn't penetrate roofs did they? More likely it was an

HE bomb with four incendiaries strapped to the fins, one of the Luftwaffe's experimental weapons. Flames were now lurching along the green-padded leather seats and oak panelling, encouraged by the keen north-easterly wind that blew up from the Thames and through the ventilating ducts in the floor. McDuell told Millar to do what he could in the Chamber, then he sent half a dozen pumps to the Star Courtyard from where they could direct water on to the Chamber above. But his priority was Westminster Hall, and that was where he would concentrate his resources.

Millar returned to the House of Commons, taking Duff with him. 'Our job was to see that the flames didn't spread across the passage and so run towards Big Ben.' Fixing a hose to one of the hydrants, the pair ran it 'through the door behind the Speaker's Chair into the blazing Chamber. Very soon there was a great rumbling ending in a crash'. A delayed action bomb from the earlier stick had exploded, sending the upper part of the west wall hurtling down into the Star Courtyard. 'There must have been thirty men in the courtyard,' reflected Duff, 'and it was filled with huge blocks of stone. It looked as though there was not more than a square foot anywhere where a man could have lived.' Somehow the only casualty among the firefighters was a man with a sprained arm.

Millar and Duff worked their way up to the floor above the Chamber, 'to find out whether the ceiling above the passage was safe. It was burning pretty fiercely and the further wall was alight'. To Duff it seemed a lost fight but Millar, who to the BBC reporter had 'all the courage you could want', disappeared and returned with a stirrup pump. 'Giving me the nozzle he pumped water to it from a bucket under the nearest tap. It seemed rather pathetic to play this little jet on to a fire of this size but it did its job.'

All the while, said Duff afterwards, 'bombs [were] still coming down and burning embers from the Commons had started fires in all other parts of the building, but the firemen got these out'. Seeing there was nothing else to do in the Chamber, Duff and Millar rejoined McDuell in New Palace Yard.

To Walter Elliot, standing on the steps of Westminster Hall at the St Stephens' entrance, the salvation of the 900-year-old roof depended on geography. The hoses that ran up the steps between his feet weren't long enough to allow the firemen to reach the seat of the fire at the far end of the roof. 'Have you a good supply of water in New Palace Yard?' Elliot asked McDuell. The Chief Superintendent told him there were two steel dams there, each holding 5,000 gallons of water fed by relays from the Thames. Elliot outlined his plan: get in through the door in the northern end of the Hall from New Palace Yard and attack the flames from directly below. It seemed so obvious but McDuell knew the old oak door was locked and he knew about Palace protocol. He didn't want to be the man who smashed it down. Elliot reassured him that he'd take responsibility. They ran down the steps at the southern end of the Hall and out along the street under the gaze of Oliver Cromwell, whose statue remained inviolate. They looked up as Big Ben's 13-ton bell struck the quarter and their spirits soared. The old boy was still alive.

If there was a key Elliot knew it would be in the key lobby next to the Commons Chamber. They didn't have the time. A fireman handed Elliot the axe that hung from his webbing belt and the Member of Parliament for Kelvingrove, Glasgow, 'had the great pleasure of laying on to the main doors of Westminster Hall – a combination of conceit and destruction which provided a certain insight into the ecstasy of the iconoclasts'. As he rained down blow after blow on the door the strong north-easterly breeze over his left shoulder sang about his ears. Elliot hesitated. Might not the 'draught sweep up the length of the Hall onwards through the blaze?' Now wasn't the time for such doubts. He drove the axe into the oak door, splintering the wood and smashing the lock. Pushing aside the main door he smashed the interior glass door with one colossal blow. 'The big heavy jets soon began to play on the blazing roof upwards instead of horizontally as formerly,' he remembered, 'and we saw the fire subside almost under our eyes. Before long it had been subdued – Westminster Hall was saved.' In fact it took over an hour to beat back the tireless flames on the roof.

Every few minutes a fireman stumbled from the Hall, drenched to the skin and blinded by sparks from the hammer-beam Hampshire Oak roof. When the battle was finally won, the dozens of firemen still inside had to wade across the flagstone floor in water up to their knees to reach the door. Elliot 'went out into the Palace Yard and there was a huge flame, waving like a scarf above Westminster Abbey'.

At 3.30 a.m. Canon Alan Don returned to check on his wife Muriel at their flat in 20 Dean's Yard. On his way he passed through a kaleidoscope of hell. All the houses in the seventeenth-century Little Cloisters 'were practically gutted'. The timber-framed Deanery, one of the finest medieval houses in London, 'was blazing and sparks were flying over the Abbey'. His wife told the Canon that four incendiaries had landed in their garden but the porter had dealt with them. 'After having a cup of tea I returned to the Abbey,' Don recalled. 'By this time flames were leaping 30 or 40 feet into the sky from the Lantern and I thought the Abbey itself was doomed. It was a terrifying spectacle.'

One of the many firewatchers positioned on the Abbey's roof estimated that at least twenty incendiaries landed from the same bread-basket. All but one were dealt with before they had time to breed fire. The one that got away punctured the exterior of the Lantern and set fire to the timbers below. When the firemen arrived the flames seen by Don had grown to 'a height of 130 feet and embers were dropping on the Abbey floor'. So too was molten lead, falling like raindrops on to the heads of the firewatchers who dashed back and forth dragging chairs and pews to safety.

The Abbey's own team of firewatchers, stationed in the eleventh-century Pyx Chapel, had done magnificent work in dousing earlier fires on the Nave and Triforium roofs, but the water had run dry. A fire on the library roof thought extinguished had in fact been lurking out of sight and now appeared with greater venom than before. More and more firemen began to arrive with water, including 'Dusty' Millar, still wearing a sheen of dust from his narrow escape outside the Commons Chamber and still with the unprepared courage of early morning, and Robin Duff with his journalist's nose for a good story. 'I went into the Abbey and it was filled with smoke,' said Duff. 'The only light came from Dusty's torch and as we went in its beam fell on to the tomb of the Unknown Warrior.' From the flames on the Lantern roof Duff could see that 'it was going to be a very hard job to check the fire. Water was needed at about 150ft above ground'.

Millar and another fireman bounded the two hundred spiral stone steps to the Triforium – the wide corridor 60ft above the floor that ran round the inside of the Abbey – for a closer inspection of the fire. In one corner of the Triforium the firewatchers had rigged up a wooden cradle capable of lowering a wounded man to the ground in an emergency. It gave Millar an idea. Returning to the ground he told a crew from Wembley to hoist a line of hose up on to the Triforium using the cradle. While the Wembley crew did as instructed, another newly arrived crew played their hoses on the south entrance to the Abbey to contain the flames threatening to reach up to the Triforium. 'The pressure [of the water] was all right,' said Duff, 'and soon great jets of water were pouring down into the Abbey through the timbers of the Lantern.'

An hour later the Lantern roof loosened itself from the Abbey's stone walls and fell 130ft. Duff heard its final groan and had time to jump clear. 'The roof crashed to the floor and the fire was out,' said Duff. The Lantern's corpse had come to rest harmlessly in the central space between the choir and the steps of the sanctuary, on the spot where almost exactly four years earlier, King George VI had married Elizabeth. The crew from Wembley, stranded 60ft up, lashed their hose to the Triforium and slid back down to safety.

Just to the south of the Abbey Reenie Carter and the girls in 'A' Division Control remained largely oblivious to the imbroglio. 'We knew there were fire engines from all over London in Parliament Square,' says Reenie, 'because our Mobilising Officer went over and came back in a few minutes saying "there's so many brass hats over there I'm better off in here".' In fact they had a clearer idea of the overall situation in Westminster than they did in their own back yard. For the past few hours they had been logging an increasing number of calls and since 2 a.m. they had been inundated. 'We hand wrote every single incident in the log books,' says Reenie, 'and there were a lot of them in Westminster that night.'

Incendiaries on Broadwick Street and Brewer Street, a major fire in Regent Street, an HE bomb at the junction of Soho Street and Oxford Street had set a gas main on fire, so too in Jermyn Street. Incendiaries on Lexington Street but no water available. The Palmolive factory in Pimlico razed to the ground. Numerous fatalities in the upmarket Dolphin Square apartments. Severe damage caused by an HE in Curzon Street, Mayfair. Shepherd Market ablaze. Samuel Pepys' House Club in Rochester Row. St John's Church in Smith Square. Bessborough Place. Claverton Street. The Fields Soap Works on West Bridge Road. And still the calls came in to the control room. Mr Ivers, a part-time ARP warden firewatching on the roof of Ganton House, reported something big barrelling through the roof of the Palladium Theatre, but no explosion. Ivers went to investigate and was joined outside the theatre by a policeman. Inside they found one of the Palladium's own firewatchers. He was ashen-faced and stammering incoherently. Finally they understood what he was saying. A 10ft parachute mine had floated down noiselessly from 15,000ft and, having gone through the roof, its 700kg of high explosive was suspended just a few feet above the stage on which Vera Lynn had sung hours earlier. Only the green silk of its canopy, caught in the Palladium's rafters, held it in check.

The biggest fire in Westminster was in Soho. Incendiaries had sent the Ship Inn up in flames at 1 a.m. and ever since there'd been a steady drizzle of incendiaries. Dean Street hit at ten past one, then Old Compton Street, Wardour Street and Bourcier Street where the Nightingales store had burned to a cinder. At 2 a.m. a breadbasket of incendiaries scattered over Dean Street and Old Compton Street. Its victims exemplified the rich diversity of the district: the Felton Advertising Agency at 58 Dean Street; the Tosca Club at No. 49 and Patisserie Valerie, pastry cooks extraordinaire, at No. 60. Budetti Luigi's hair salon at 41 Old Compton Street; Café Belge at No. 56, the Algerian Coffee Store at No. 52 and the Minella & De Rossi Café Bar at No. 61. At 2.43 a.m. Reenie logged a call from Dean Street: 'AFS urgently reqd. Fire spreading.'

The call came from the handful of Reserve Policemen in Dean Street, one of whom was the recently arrived Ballard Berkeley. After the initial flurry of incendiaries on his beat in the Haymarket it had been a relatively quiet night for the 36-year-old actor. 'So I went down with a friend of mine who was in Gerrard Street and there was a building going down there.' Ballard helped evacuate people and then did a sweep of the neighbouring buildings to make sure there were no incendiaries burning unseen. From the roof of one he looked north and 'saw a tremendous blaze at Old Compton Street and Dean Street'.

It took Ballard only a few minutes to get down from the roof and across Shaftesbury Avenue to the conflagration. The fire he had seen came from the wholesale bedding manufacturers at 58 Old Compton Street. 'It really was ablaze,' he recalled, 'and the mattresses were lovely fodder for the flames.' To Ballard, 'it looked as though the north side of Old Compton Street from Dean St to Oxford St was going to be burned down because the fire service was so extended it couldn't possibly cope with all the fires'. Having put in the call to Reenie Carter in the control room at 2.43 a.m. the police waited for the sound of fire engines jolting along the rubble-strewn streets. They had no idea that nearly all available fire crews in Westminster were tackling the blaze at Westminster Palace. By 3 a.m., with no sign of a fire engine, the police decided to evacuate Old Compton Street, its people and any perishable food stuffs from the shops. Ballard and his colleagues went from house to house ordering people to leave. When that was completed the police and the shop owners began emptying their premises of contents. Everything was piled neatly outside on the pavement and within minutes a sordid gaggle of spivs and chancers 'were trying to buy the stuff cheaply off the people who owned the shops'.

An off-duty Canadian soldier, 6ft 4in and ready for anything, latched on to Ballard. 'He kept calling me "captain",' remembered Ballard. '"What can I do, captain?", "Can I help you, captain?" And he was as drunk as a drunk could be.' From time to time, when he thought no one was looking, the Canadian reached into the pocket of his greatcoat, pulled out a bottle of whisky and drank greedily. 'I was trying to shake him off but he insisted on coming down Old Compton with me and I couldn't get rid of him.' The massive Canadian soon proved his worth, 'picking up great things and taking them out' from the fire-damaged shops. Ballard left the Canadian to it and moved further along Old Compton Street doing what he could to help the distraught shopkeepers. 'Suddenly I saw a chap fly out of a door, one of our chaps,' recalled Ballard. 'I picked him up and said to him "what are you doing, flying out of doors?" He said "My God, I've really been hit". "Hit?" I said. "Yes, some Canadian bastard in there is looting all over the place".' The Canadian soldier had emerged from the shop and now filled the doorway with his alarming physique. Even from a dozen yards Ballard could see his greatcoat was 'full of cheese and wines and whisky and cigarettes'. The sergeant in charge ordered his men to 'get the fellow out of there', and then disappeared rapidly on another job. 'We didn't know how to tackle him because by this time he was raving drunk and a tremendously strong man.' Ballard lured the drunk out of the doorway with the promise of more rich pickings and two policemen 'came up behind him and put a cloth over his head. As he went to get the cloth off they got both his arms and told him they'd break them if he didn't give in'.

The threat had a sobering effect on the Canadian. He went limp and then became maudlin, mumbling that he was a worthless no-good and begging the policemen to forgive him. 'We had no time to arrest anybody,' said Ballard, 'so I had a chat with him and said "come on, you're helping no one but the Germans by looting. Just go on your way".'

It was now just after 3.30 a.m. and the threat of the fire spreading had diminished. With the drunk Canadian out of the way Ballard could get his breath back. Then he felt a hand on his shoulder. Spinning round he saw the grinning face of George Niblett, a great pal of pre-war days and now a fellow policeman. '"Is this your beat?" he said to me. "No, I've just come down to help". So had George.' The two had a chinwag, caught up with each other's news and parted with a promise of a pint in the not-too-distant-future. 'I'd only just left him,' remembered Ballard, 'when suddenly it happened. It's the most extraordinary thing when it happens.'

Ballard had just plucked a case of drink from a shop and put it on the pavement opposite. He walked back across the road and 'suddenly everything stopped and there was complete and utter silence. It was so quiet it was unbelievable'. Like everyone else, Ballard had been told that you never hear the bomb with your name on it. It just explodes on top of you. 'I heard no bomb or no explosion but whether the explosion had deafened the eardrums I couldn't tell you,' he reflected. 'I wouldn't know.'

In the seconds after the bomb landed on top of Harris & Co, Hatters and Hosiery, 43–45 Old Compton Street at 3.45 a.m., 'everything was silent and still'. The chin-strap of Ballard's helmet had nearly throttled him as the pressure of the blast tried unsuccessfully to rip it from his head. Now he lay on the ground, blinking with large uncomprehending eyes. 'The whole place was filled with grey mist, it was like a theatrical scene ... Really quite beautiful.' For a moment he wondered if he was back on stage but then the bricks and glass began to rain down. 'I saw somebody and tried to move but couldn't and thought "God, my legs have gone".' Ballard looked down and realized his legs were pinioned by rubble. Slowly and methodically, without flapping, he removed the debris and got to his feet. And then he was overwhelmed by a 'tremendous panic'. 'I was frightened and scared and I had to get away. I had to run, I must get away, I must run. And I did, I ran.'

Ballard, stumbled, reeled, staggered down Old Compton Street, tripping over chunks of masonry, slipping on puddles of blood, past the shattered corpse of George Niblett. Nothing mattered, just as long he got away. 'And then I stopped, absolutely dead. And I stopped because I was no longer there, I was outside my body ... I was actually outside my body and looking down at myself and talking to myself and saying "Why are you running and why are you frightened? This is your job, you're supposed to do something here", and then I seemed to enter my body again and I was calm and cool and collected and it was the most extraordinary sensation I've ever had.'

Ballard turned and walked steadily back down Old Compton Street. 'I was still wearing my helmet so I was certainly not concussed, possibly shocked and I was cut and bruised but I was in one piece.'

He looked down at Niblett and thought what 'an awfully nice chap' he'd been. For the next minutes Ballard dug out the wounded and helped the shocked to a warden's post and made sure they had a nice cup of tea. 'We thought that cured everything.' Another policeman took Ballard by the arm and led him to a man holding on to life by his fingertips. 'There was my Canadian friend lying there. He looked just like a giant rag doll.' The drunk was quite peaceful; no anguished screams, of 'Why me?' No self-pity. 'Hello captain.' He smiled at Ballard. His back was broken. 'You all right?' Ballard smiled back. 'I'm all right, captain. Can you give me a cigarette?' Ballard choked. He thought of parts he'd played in a previous life, with a dying man requesting a last cigarette before the curtain fell. 'This doesn't happen in real life?' he thought. Ballard knelt down by the Canadian and put a cigarette gently between lips coated grey with plaster dust. He lit the cigarette. The Canadian took a long, deep, contented drag, 'Thank you, captain'.

DEATH had become almost mundane by 4 a.m., the peak of the second wave of bombers. The dead lost their individuality and became just figures on an ARP's casualty chart: 'five dead here', 'seven killed there', 'dozens over the road'. Only occasionally was the horror of death sufficient to create a shock. One man recalled seeing a 'dog trotting down the road with what I thought was a bone in its mouth; it was in fact a baby's arm'.

Those still alive huddled in their shelters or in their houses, their faces given a pewter sheen by fear and their eyes dark and sunken by exhaustion. Silently they prayed that the bombs would fall somewhere else, anywhere else, just not on them. 'We were all guilty of sheer relief when we realised that they had not hit us,' Joan Veazey says. 'Someone once wrote in one of the papers that a true Christian would pray that the bomb would fall on them rather than on their fellow man. I'm ashamed to think that I'm not a Christian but just a human, who is very scared.'

Death made no distinction between the pious or the impious, the reverent or the irreverent, the famous or the ordinary, the rich or the poor. If your name was on a bomb, you died. The Reverend Stanley Tolley, vicar of St Silas in Nunhead, never heard the projectile that killed him and the six air raid wardens he was chatting to outside his church; the bomb that crashed through the roof of 43 York Terrace, Regent's Park, exploded as the Sacrifice and Service sect worshipped the moon. Among the many dead was the sect's leader, a 54-year-old Dutch doctor called Bertha Orton. The large diamond-encrusted gold cross hanging from her neck was removed by police before looters got to it.

Alfred Atkins was killed firewatching in Streatham a few hours after he had appeared on stage in the West End's Wyndham's Theatre in *Cottage to Let*. He and the rest of the cast were about to star in a film version of the wartime spy thriller.

Eighteen-year-old the Honourable June Forbes-Sempill, engaged the previous month to a pilot officer in the RAF, walked out of her house in Basil Street, at the back of Harrods in Knightsbridge, and climbed behind the wheel of her WVS canteen van. She had driven 100yds when a high explosive whistled down and disintegrated the vehicle.

Bob Ellis was an ordinary husband and an ordinary father; though like every father he was considered extraordinary by his daughter. Joan had married a few months earlier; Edward Wallstab, an intelligent young man with good prospects. His parents were German and Bob Ellis would have preferred his daughter to marry 'a nice English boy', but he said nothing and he didn't interfere. Eddie was called up in early 1941 and Joan moved back in with her parents in their flat above the County of London Sessions House in Newington Causeway where Bob was a custodian.

Joan and Emily Macfarlane had been best friends for years. They worked together for Hardings, the hardware company, before the war and many an evening they had spent at the Trocadero watching the shows and flirting harmlessly with the boys. Now Joan was married she missed her husband, and each day she cried a little bit more. 'My father asked me "What would make you happy?", and I told him I wanted to go and see Eddie at his camp on the south coast,' remembers Joan. 'So on the Friday afternoon [May 9th] he took me down to Victoria coach station, bought me a return fare and gave me spending money for the weekend.' As the coach departed Joan looked back at her father from the rear window. He blew her a kiss and smiled.

Thirty-six hours later Bob Ellis was on firewatching duty at the Sessions House when he heard an unusual noise, something flapping on the recreation ground at the rear of the building. Ellis wasn't a man given to nervous wavering. Fifteen years earlier, as a detective, he'd been awarded the King's Police Medal for subduing an escaped convict who'd threatened him with a pistol. Now he strode off to investigate. The parachute mine exploded a minute later. The person who logged the incident in the Southwark control room noted that 'a building of 3 floors about 200 by 100ft used as lecture rooms and offices and contents severely damaged by explosion'.

THOUGH the eye of the storm had passed, the Luftwaffe hadn't yet finished with Westminster. Five minutes after the bomb on Old Compton Street – later identified as a 1,000kg thin-cased 'Hermann' – a high explosive of a similar size exploded in the three-storey electricity bulk supply station of the Central London Electricity Company in St Martin's Lane. The girders that supported the upper floors bowed under the weight of 2,000 tons of machinery and large quantities of sulphuric acid. For half an hour the girders sagged. Then they buckled and the upper stories cascaded down on to the basement where thousands of gallons of oil were stored for diesel engines. Henry Holland, the power station's 66-year-old caretaker, and his wife Florence, already trapped in their basement shelter, were flattened in the crush. A fireman described how 'the mass of oil and acid and debris boiled through the night in a dark and hidden inferno of twisted girders and huge iron automata'.

In Mayfair, bombs of varying description had been dropping on Shepherd Market since two minutes past midnight. It had been terribly exciting at first for Madeleine Henrey and her Old Etonian husband, Robert, camped in the corner of their flat with their baby between them, listening to incendiaries clattering on to the cobbled streets of Shepherd Market. They had 'done their bit' chasing incendiaries, together with the wardens and firewatchers and other residents. Everyone was pulling their weight, remembered Madeleine, even Shepherd Market's prostitutes, whose towering heels caught in the cobbles as 'still with fur coats [they] ran hither and thither with pitchers of waters'. An incendiary landed on a draper's store and soon wardens were climbing in through the smashed shop front to salvage what they could. 'Dozens of boxes of silk stockings and newly delivered lingerie were piled under an archway,' recalled Madeleine, 'but as courtesans returned with their pitchers the boxes disappeared as if by magic.' Soon the fire had spread to Sunderland House, a grandiose building all crystal chandeliers and Chippendale furniture, built by the American industrialist William K. Vanderbilt as a wedding present to his daughter when she married the Duke of Marlborough. As the flames swarmed over the House a firewatcher howled in helpless fury, 'This building is doomed! It was built on consecrated ground, on the site of May Fair Chapel!' Madeleine glared at the man with French disdain. 'This philosophical outburst in the face of so much danger staggered me, for it was clear that soon the building would become a raging blaze.' Seeing this particular fight was lost Madeleine returned to check on her baby. As she did she 'saw a young woman, Winifred Berry, who lived in our block of flats calmly painting the scene from her sitting-room window'.

The second wave of German aircraft brought high explosives for Mayfair. Madeleine and her husband sheltered in their flat. 'Each time a bomb whistled down we both rose hastily to throw ourselves over the baby as if our bodies could save him from the explosion,' said Madeleine. 'But our haste was such that we invariably knocked our heads together on the way up, relieving our fear with laughter.'

At 4.16 a.m. a bomb landed on Curzon Street House, a modern block of steel-framed flats hit badly in a previous raid. A few hours earlier it had been possible to reach Curzon Street from Shepherd Market by a captivating covered passage. On one side of the thoroughfare was a court in which a weekly flower market was held. Opposite was Paillard's, the most celebrated hair stylist in Mayfair. Paillard himself was an elderly dapper Frenchman, not lacking in self-regard. His name had been written in large gold letters above his shop window in which were displayed the tortoiseshell combs he had used when he opened the shop forty years earlier. But between midnight and 4.16 a.m., fire had fed on his lacquers and shampoos and his tortoiseshell combs.

23

BY 4.30 a.m. the storm had swept over Stepney leaving a trail of destruction in its wake. In the control room of 22 Station in Burdett Road, Florrie Jenkins was matted with dust 'that had come through the ventilation shaft of the control room'. Throughout the night she and the other firewomen had remained at their desks logging calls against a background of violent thuds, some more distant than others. 'A couple of times bombs dropped so close the whole building shook and the discs on the board that represented our engines fell on to the floor.' Florrie picked up the discs and tried to remember which number went on which hook. She returned to her desk and told herself her mum was fine, probably curled up in her bed in Coutts Road, just round the corner from the fire station. Then shortly before dawn her Mobilising Officer, Mr Nash, took Florrie to one side. 'I think you better pop home,' he told her, 'some stuff has come down on Coutts Road.' Florrie flew out of the station. The blue police box on the corner had taken a direct hit and several of the shops on the other side of the road were gutted by fire. Florrie started up Burdett Road but at the top a police cordon barred her way. 'Sorry, love,' a policeman told her, 'there's a UXB in Coutts Road.' 'So I went back down Burdett Road and tried to get to my house that way,' says Florrie. 'But there was another cordon at the bottom. By this time I couldn't care less about the police so I went under the cordon and into Coutts Road. The air raid wardens were doing their nut chasing after me but I just pointed and said "That's my house!".'

Florrie ran up Coutts Road, her feet crunching on the glass blasted from the windows of the hairdresser's opposite her house. A bomb had landed right on top of the hairdresser's, decapitating the owner asleep upstairs and 'removing the front of my house'. The wardens caught up with Florrie as she stood staring into her front room. 'Where's my mum?' she asked. 'Your mum's OK,' one of the wardens said, 'we got everyone out safely.'

Dr Kenneth Sinclair-Loutit had been sent to Stepney a couple of hours after his attempt to save the coffee warehouse in the City. He and his Heavy Rescue team were now trying to extricate survivors from a collapsed block of flats. Early on in the Blitz Sinclair-Loutit had visited his wife's family in Abertillery and gleaned valuable information from the miners among them about cave-ins and how to spot the early signs. 'As a good luck present they gave me a miner's pick that went with me during air raid duties,' he recalled. 'It had saved my life once already during the Blitz when the tunnel I was in began to give way. I used the blades as a short roof beam and managed to halt the cave-in.'

Now he was doing what he hated most, carrying out a rescue while a raid was in progress. 'I can't remember where exactly in Stepney the flats were,' he says. 'On 10 May the borough was hit so badly it was just a jungle of smoke and flames. I led my rescue team into the wreckage and the first few yards of tunnelling were always the worst; if the building was going to cave in on top of you it would most likely be at the start.' Every bomb that dropped, he says, was 'a form of Russian Roulette in which the trigger is pulled by someone else … the noise of descending bombs, with the shake of the ground under your belly while tunnelling underground to dig out those already imprisoned by earlier bombs, leaves you with nothing to imagine'.

Sinclair-Loutit found an elderly couple, alive but in a state of severe shock. He extricated them using the 'wheelbarrow technique', the most favoured method of bringing out people from a low tunnel in which the rescuer straddled the victim who was on their back. The trapped person's hands were tied together and their arms slipped over the neck of the rescuer. He then propelled them both along the tunnel using his hands. 'The wheelbarrow technique was the most effective,' recalls Sinclair-Loutit, 'just as long as the person you were rescuing wasn't hysterical and liable to try and strangle you! On this occasion the old couple were very brave. I remember the man joking to me as I put his arms round my neck. "Don't you try any funny business", he said to me.'

Most Civil Defenders had nothing but respect for the men in Heavy Rescue; they considered it a job for those either too dense to imagine the ways they might die, or for those more courageous than themselves. Their fortitude and stoicism rarely wavered throughout the hours of rescue work. Only once did Sinclair-Loutit remember seeing one of his rescue team show any irritation. After bringing out several dead from a building one of them walked over to a WVS canteen van. The ladies serving could offer him only sardine sandwiches. 'Have you got anything else, love?' he snapped. 'No, sorry, sardine only. What's wrong with sardines?' The man looked up at the smartly-dressed woman smiling behind the counter. 'Nothing, just that sardines smell like dead people.'

BY 4.45 a.m. the north-easterly wind that had galvanized the fire in the Commons Chamber had shifted direction. It was now blowing east, a useful tailwind for the Luftwaffe bombers heading home to France having dropped their load. In the City of London the stench of raw sewage from fractured pipes wafted along Queen Victoria Street, a thoroughfare opened in 1871 to ease the west-bound traffic from Cheapside, and which ran from Mansion House down to the neo-Romanesque red granite piers of Blackfriars Bridge. Here the situation was dire. The incendiaries dropped by Hufenreuter and his crew on Cannon Street Station around midnight had started a conflagration now threatening St Paul's Cathedral to the north and the whole of Queen Victoria Street. Bombs of varying description played havoc in the eastern end of the Street: Mansion House, the 200-year-old official residence of the Lord Mayor of London, where once Emmeline Pankhurst had been imprisoned, had lost almost all its windows, most notably the splendid stained glass in the Egyptian Hall. Some of the Portland stone walls had been defaced by shrapnel, too, but an army of firewatchers now swarmed over the 23,000 sq ft of Mansion House, dousing incendiaries with their stirrup pumps. Close by, St Stephen's Walbrook, a Wren church ravaged six months earlier, had been given the *coup de grace*. The disfigurement continued all the way down Queen Victoria Street to the intersection with Queen Street, where the flames had destroyed the art-deco Chamber of Commerce as well as Sweetings, the famous City oyster bar on the corner with its Venetian Gothic style and intricate facades. Nearby, the one-roomed Sugar Loaf pub, which had seen its fair share of raucous City drinkers, succumbed to the flames.

The fires now gambolled down the north side of Queen Victoria Street and, with no respect for history, consumed St Nicholas Cole Abbey and its 135ft spire. Christopher Wren had rebuilt the church after its destruction in the Great Fire of London in 1666 for £5,042 6s 11d. Included in this cost was £2 14s, 'dinner for Dr Wren and other company', and 'half a pint of canary (sweet wine) for Dr Wren's coachmen – 6d'.

Next to go was a Barclays Bank, a tobacconist and a shop that specialized in burglar alarms. Opposite, on the south side of Queen

Victoria Street at No. 101, stood the International Headquarters (IHQ) of the Salvation Army. Next door at No. 103 was the Army's Assurance Society. Incendiaries had fallen on the roof of No. 103 at 2 a.m. but Major Kennard, the Army's HQ housekeeper, had put them out with the help of a firewatcher. Captain Wesley Grottick from the Army arrived not long after, having collected a grey canteen van from their Barking base. He and two female Salvationists found 'Queen Victoria Street and the surrounding buildings devastated but IHQ still standing intact'. Grottick replenished the canteen van with sandwiches and hot drinks from the kitchen of IHQ as 'the best I could do was keep the firemen, public and other, emergency workers supplied with tea and such provisions. Added to the numbers were the telephone staff from Faraday House Exchange who had been on duty all night during the raid'.

Faraday House was less than 100yds down from the Salvation Army HQ at the westernmost end of Queen Victoria Street. Inside the ugly white concrete behemoth, rising nine-storeys high and sprawling over 10 acres, was the world's biggest telephone exchange, keeping Britain in contact with the outside world. There was also the Overseas radio link with America and the Commonwealth and, in the north-eastern block there was what was known as The Citadel, an emergency bunker for the Cabinet. Churchill had rooms here under the 7ft-thick reinforced concrete roof, but tonight the only two members of the Cabinet in residence were Sir John Anderson, Lord Privy Seal and the man who'd given his name to the air raid shelter, and Ernest Bevin, Minister of Labour. For Fire Control in Lambeth the wellbeing of the two politicians was of little importance. What mattered was saving the telephone exchange. Without it Britain was cut off from the rest of the world.

Back up Queen Victoria Street the flames were coming closer, leaping up and down, up and down, like a line of men doing physical jerks. The offices of Chubb & Sons, the locksmiths, were engulfed, then the Post Office next door and an ABC café where during the day customers could drink their tea over a game of chess. Safe on the south side of Queen Victoria Street Major Kennard and his wife watched from their bedroom in IHQ as the flames advanced. They could see the firemen trying to establish a firebreak in 134 Great Victoria Street, an office block destroyed in a pre-war fire in 1939. It was the last chance to save the buildings beyond: after No. 134 it was the red brick College of Arms, in which was housed the signature of every English monarch stretching back to William the Conqueror, then it was the narrow Godliman Street, which ran uphill 100yds north to St Paul's Cathedral, whose dome looked like an island in a raging orange sea, and whose eastern walls had been nibbled at by high explosive shrapnel.

On the other side of Godliman Street was the east corner of Faraday House where some firemen, those whose sense of humour was still intact, pointed out to one another the words above the main entrance: 'The Word of the Lord Endureth Forever'.

In the HQ of the Salvation Army Major Kennard was alerted by a shout from the nightwatchman at around 4.30 a.m. The shifting wind had blown flames from fire behind him, in a paper mill in Upper Thames Street that ran parallel to the river, on to the back of the Salvation Army Headquarters. The fire spread so quickly through the Assurance Society building that Major Birbeck the caretaker didn't even have time to retrieve his spectacles. Trapped inside somewhere was 'Ginger', the Army cat who controlled IHQ's rat population. Kennard and his wife got out of IHQ in just the clothes they were wearing. One Salvation worker rushed out of the building, tripped over a fallen phone wire and was run over by a fire engine tearing down Queen Victoria Street towards Faraday House. All the while from inside his canteen van Captain Wesley Grottick watched 'the gradual destruction' of IHQ.

Down by the shell of 134 Great Victoria Street fireman Nobby Clark and his crew had just arrived after taking three hours to cover the 30 miles from Hadleigh in Essex. They laid their three hoses in a 5,000-gallon static dam fed by a relay in King William Street and started to fight the flames on the ABC café. The heat was so great the water turned to steam as soon as it left their hoses. Clark soon 'felt dazed and giddy'. Turning his head to one side he retched, his vomit splashing the baking pavement with a violent sizzle. In front of him through the clouds of steam he could see 'a turntable ladder working at about 100ft at the far end of the Salvation Army building which was well alight'. After a few more minutes Clark found himself 'unable to think and [I] felt the strange urge to walk towards the heat'. Instead he gripped the branch even tighter, leaned forward through the smoke and dust and sucked what little air there was coming from the nozzle.

The fires in the ABC café were being checked when the hoses of Clark's crew and their neighbours failed. Sending a couple of men to see what the problem was, Clark opened up a large hydrant nearby. Dry.

Captain Wesley Grottick and his two female Salvationists heard from the firemen queuing patiently for tea that there was not 'sufficient water as the mains were broken and although they tried to get it straight from the Thames the tide was so low the mud choked the hoses'. The firemen waiting in line for refreshment shuffled wearily towards the counter. Only when they realized they were being served by women did they straighten their backs and jut out their jaws, deriving inspiration from such disregard for danger. George Woodhouse, up from a sub-station in Holloway, never forgot the 'two Salvation Army lasses who were handing out cups of tea and a biscuit as if it was a Sunday school picnic. They appeared to be completely unaware of the bombs falling around them'. Throughout the Blitz Woodhouse had 'often wondered what the word courage meant, but on that night those two lasses had it in abundance'.

Woodhouse's thoughts were stolen by an ungodly whoosh. So were Nobby Clark's. He looked up as 'the whole of the burning Salvation Army building collapsed into the street, completely blocking the road'. The turntable ladder was still erect, a beacon in a storm that was now almost uncontrollable. 'Many firemen ran towards the scene and started getting injured firefighters out of the chaos,' recalled Clark.

It was now 5 a.m. and London was confronted with its gravest crisis of an already long and tumultuous night. Major Frank Jackson, Chief Officer of the London Fire Service, stood impassively at the western end of Queen Victoria Street as his men were dragged clear of the turmoil. It hadn't been a lucky night so far where water was concerned; just after midnight a bomb at Cannon Street had fractured a major City water main that connected to the Grand Junction Canal. Around the same time the West End main, from the Regent's Canal in Regent's Park to Shaftesbury Avenue, had been blasted. Both were capable of reinforcing public mains at a rate of 30,000 gallons of water a minute. Then at 3 a.m. an even bigger catastrophe: three HEs had fallen on Brecknock Road in Tufnell Park rupturing the two 36in trunk mains that supplied the City from Crouch Hill Reservoir in north London. What with the damage to the mains and the unusually low tide in the Thames, water had been hard to come by.

An hour earlier Jackson had instructed Chief Officer Arthur Johnstone to organize a fleet of lorries to bring water to Queen Victoria Street from the Regent's Canal, a mile north. Each lorry, winding its back slowly through streets blocked with debris or craters, tipped 1,000 gallons into the 5,000-gallon steel dam. Until the Thames started to rise the lorry relay was the best hope of saving Faraday House. If the relay failed and the fires began to cross Godliman Street, Jackson had orders to dynamite the street to cause a natural firebreak. He knew it was a last resort; such an explosion might jeopardize the foundations of St Paul's Cathedral 100yds north.

24

'AT about 5 a.m. I looked across the river and saw that Lambeth Palace was on fire,' wrote Canon Alan Don in his diary, 'so I crossed the bridge to see what had been happening.'

It took Don a little over ten minutes to race over the bridge and up Lambeth Palace Road. He barely noticed the fires in St Thomas's Hospital but inside the situation was critical. 'Masses of hot debris fell from time to time down the lift shaft to the basement [where Casualty was],' recalled Sister Casualty Annie Beale, 'and to make matters worse, a large pipe burst and there was a flood and a loss of precious water. This was squeegeed [sic] away and swept away from the patients in the corridor.' Eighty people had been brought in to the St Thomas's Casualty Department during the night, most suffering from minor wounds, but fifteen had been admitted to a temporary ward in the basement. 'For about two-thirds of the time I was semi-conscious under sedation,' a wounded fireman remembered. 'During a period of awareness I realised the basement was many inches deep in water and I could hear a trailer pump at work not all that far away. In response to my questions I was told that "we have a burst water pipe and the fire brigade are pumping out the flood for us".' The wounded man lay back in his bed, 'able to feel some very near misses', but gradually succumbed to wisps of unconsciousness drifting through his mind.

Canon Don hurried through the red-brick Tudor gatehouse that led into the grounds of Lambeth Palace. His first question concerned the Archbishop. Was he safe? Chaplain Ian White Thomason told him that Cosmo Lang had retired to his bed at 3 a.m. after being blown off his feet by a bomb blast. In the two hours since the Archbishop had been heard 'to complain that he couldn't get to sleep with the noise that was being made'. There was little one could do about the Luftwaffe, replied Don. Thomason shook his head mischievously. It wasn't the 'bombs and gunfire' disturbing the Archbishop's rest 'but the crying of a baby in the passage outside his room'.

Don scanned the Palace buildings, assessing the extent of the damage. The fire service had taken an eternity to arrive, he was told, and though they now had the fires under control, 'nothing could save the chapel roof and the top storeys of Lollards' Tower'.

Thomason told Don that when the roof of the chapel had collapsed, 'some damage was done to the Laudian Screen and the Stalls'. Don could see, too, that 'the roof of the library had also been burned about halfway across and the whole place was an appalling mess'. Thomason and his staff had worked tirelessly throughout the night to remove books and manuscripts, but he estimated that many hundreds had been lost to the flames.

AT Waterloo Station the asphalt on the platforms was soft and spongy from the heat of the fire in the 23 acres of vaults underneath. Mac Young and his crew from Paddington had spent much of the night fighting the fires in the vaults assisted by ten pumps from outside London. 'These crews turned up and they all had their epaulettes on and looked very smart indeed,' recalls Mac. 'They also had these lovely turntable ladders on their pumps which could be elevated and extended.' Mac stood by his pump, watching, a cigarette wedged between his index and middle fingers, as one of the provincial firemen climbed towards a platform at the top of the first section of the 100ft ladder. 'He got on to this platform where there was a fixed hose and clipped himself on,' remembers Mac, 'and signalled to the operator to elevate him up. But in the excitement of the moment he forgot to keep his toes clear of the edge of the platform and as soon as the platform started rising his feet got caught in the first rung of the ladder, breaking all his toes.' Despite the accident the fireman remained on the platform as he rose up towards the fire on the first floor of the office block. 'He was a brave chap, this fireman,' recalls Mac, 'but within a few minutes he'd passed out through a combination of heat and pain from his broken toes.' The problem now was how to get the unconscious fireman down from the platform. 'The turntable operator couldn't risk bringing the ladder down automatically,' says Mac, 'because the fireman was slumped across the platform and one of his arms or legs would be trapped in the rungs.'

Mac caught the panic rising in the provincial crew. 'I told the operator to swing the turntable out away from the flames and I climbed the ladder to bring him down.' The platform, remembers Mac, was 'awash with blood from this poor man's broken toes' as he unclipped the fireman. 'I'd been trained in bringing down a 12-stone man from a burning building so I just hoped he wouldn't be more than that.' Mac hoisted him over his shoulders and descended the ladder, one rung at a time, congratulating himself for having replaced the standard issue rubber boots with a pair of leather ones he'd wangled from a pal.

While Mac and the ten pumps tried to contain the fires in the office block in York Road, fires in the vaults on the other side of Waterloo Station were going berserk. An HE with incendiaries strapped to its fins had exploded in the electric light department, blasting down the door that opened on to Lower Road, which ran parallel to platform 1. Like shoppers jostling to get through the door on the first day of the sales, the flames thrust and shoved their way out on the streets and climbed over the parcel vans and motor horse boxes parked outside. Then they started barging their way into other vaults.

By 5.15 a.m. the fires at Waterloo Station stretched from the offices on York Road on the north side, right underneath the platforms and concourse to the arches on the southern side by Lower Road. The people sheltering in the cavernous public lavatories were safe, but the heat was stifling and with no electric light or power each of them prayed for the sound of the All Clear.

Opposite Waterloo in The Cut, Fred Cockett's hose-laying lorry had 'saved our baker's from destruction'. The fires had been tamed but the eastern end of the market street was black and charred, and still hissing as the firemen played water on the ruins. The Linoleum warehouse had been partially saved because of its structure. 'It was an old warehouse,' remembers Fred, 'with timber frames a foot square that took a long time to burn, so we had time to get them under control. They were much easier than the steel-framed buildings because there the heat quickly contorted the steel. We'd had men injured by girders that expanded with the heat and smashed through walls on to them.'

Round the corner from The Cut in Waterloo Road the Davey Gregg food emporium was a forlorn sight. The counters that ran down each side of the store looked like rotting black stumps of teeth and only the legs remained of the chairs put out for worn-out shoppers to take the weight off their feet while they waited their turn. And the produce was ruined, all the ham and butter and bacon no longer edible after gallons of dirty water from the Thames had been used to quell the fires.

Hose relays ran south from the Thames by Waterloo Bridge, up past the station, past The Cut, past Davey Gregg's food emporium, over St George's Circus where the hoses weaved round the hideous remnants of three fire engines destroyed in the Surrey Music Hall explosion, and down to the still raging fires at the Elephant & Castle. Then the line went dry and Divisional Officer Geoffrey Blackstone's temper snapped. He jumped in his staff car and followed the relay north until he found the cause of the interruption: another officer had diverted the water from the relay to confront his own fire. Blackstone had his water back within minutes.

The emergency steel piping, laid in 20ft lengths from Westminster and Waterloo bridges, had rescued the Elephant from annihilation, but at 5.30 a.m. the fires had yet to lose their appetite. From her canteen van under the railway bridge in the New Kent Road, Emily Macfarlane watched as the Elephant she had grown up in, shopped in, danced in, melted before her eyes. 'A big tea warehouse in Tarn Street near to the Trocadero caught fire,' she recalls. She dashed forward, 'but every time I went towards the Elephant and Castle I was told "this is no place for a woman" so I went back to the canteen van and carried on dishing out cups of tea'.

But for the last hour she had been edging closer to the conflagration, desperate to help the firemen collapsing all around her. Those that had been at the Elephant since midnight fought the fires with their hoses wedged into the crook of their elbows, their other arm over the top, controlling the jet. Staying awake was a constant struggle and sometimes they lost, surrendering to tiredness. The moment a head drooped the hose broke free, bucking and kicking like a runaway horse. Elsewhere in the Elephant, on the pavement in Newington Causeway, outside the mercifully spared Elephant & Castle pub or by the railway booking office, firemen sat resting for five minutes, a cup of tea or cocoa cradled in one hand and the other supporting the weight of their lolling head.

At 4 a.m. the yardmaster at the Elephant & Castle Underground station had decided that the finances of his employers took precedence over his own welfare. 'I began to get a bit worried about the Board's money,' he recalled, 'which was in the booking-office safe, and also about a lot of cloakroom deposits, as the walls of the booking-office and the roof were being seriously threatened.'

Together with another employee, the yardmaster 'emptied the safe and took the books from the office and put it all under lock and key downstairs'. He put the station's float and the cloakroom deposit money in a disused lift and then 'thought about the other station* across the road, which is also under my control'. He and his companion took a step outside the station offices. 'Things were pretty hot over there,' the yardmaster noted, 'one seething mass of fire all around.' The pair looked at each other, managed an empty smile, and then sprinted through the cauldron towards the other station. 'I decided to do the same with this booking-office money as I had done on the Bakerloo side ... put it under lock and key.' That done, they made to leave, when 'all of a sudden another bomb dropped and the blast blew doors and windows out and sent us ducking for cover'.

The yardmaster and his colleague helped each other up, brushing the dust from their uniforms with the back of their hands. They didn't know it, but one of the last bombs had just fallen on London. Since 5 a.m. the Luftwaffe downpour had begun to ease. At 5.37 a.m. the north-west turret of New Scotland Yard was hit by a small high explosive bomb. Then quiet. The All Clear went at 5.52 a.m. For one minute the siren blasted, steady and joyous like a liner arriving in port, so different to Londoners than the awful banshee wail a few hours earlier. Different, at least, to those who were still alive. Emily Macfarlane defied instructions and walked towards the Elephant & Castle. It was a terrestrial hell, and she knew in that moment it had gone forever, no matter how bravely the firemen stood up to the bullying

* There were two entrances to the Elephant & Castle Underground station; the Bakerloo entrance at the corner of London Road and Newington Causeway, and the Northern Line opposite the Tabernacle.

flames. A man walked towards her, an anonymous black figure against an entrancing orange backdrop. 'It was an old mate of my dad's,' recalls Emily. 'They'd fought in the Boer War together and usually he was very cheerful. He'd always give you a hug and ask after the family.' His face was haggard. There was no smile. 'He put his hand out to stop me going any further and told me there was nothing I could do up ahead. It was too late.'

25

BY 6 a.m. the carcass of Albert Hufenreuter's Heinkel He111 was covered in a fur of frost, guarded by soldiers from the local Home Guard platoon. They blew into their hands to warm them as the lavender blue dawn broke over Kent. Thirty miles north at West Malling airfield all of squadrons 29, 74 and 264 had returned safely. 'The thirty-two aircraft had all landed now,' recalled Flying Officer Freddie Sutton, 'and between the lot had shot down sixteen Huns ... and I think there is no shadow of a doubt that many others besides were so severely mauled that they never reached their home bases.' Other such exuberant claims were being made at airfields across southern England.

At Hunsdown aerodrome in Hertfordshire, three Douglas Havoc crews from A Flight 85 Squadron made out bullish reports. A Ju88 and a Heinkel He111 had been 'damaged' in two attacks while Canadian Flight Lieutenant Gordon Raphael described how his radar operator Bill Addison had guided him on to the tail of German aircraft over north London: 'I closed to 100yds and recognised e/a as an He111. I fired one burst of 4 seconds at the starboard engine and e/a immediately burst into flames and went into a left hand spiral dive to approximately 8000ft when it levelled to a 60° dive to the ground. It appeared to explode just before it hit the ground.'

At Northholt in west London a 33-year-old Hurricane pilot with 306 Squadron had been credited with a Heinkel He111. Flying Officer Wladyslaw Nowak was the first Polish day fighter to have notched up a night kill. He was jubilant because the war for Nowak was personal. He had been captured by the Germans in September 1939 and sent to a POW camp in Romania. From there he'd escaped and made his way to France, then on to England and the Royal Air Force. Nowak had had no news of his wife for nearly two years; she had joined the Polish Underground when Germany invaded, so he guessed she was dead.

Nowak was only too happy to wax lyrical about his night's work to the squadron's diarist. 'It was a fine warm evening,' he told him. 'The sun had just set and the first star was already gleaming when the first AA shell burst over the city. A raid was on. We took off one after another at five-minute intervals. It was pitch dark in the west but London was aglow with the explosion of bombs, AA fire and burning buildings. I called Ops [Operations] and got my course. As I climbed and then levelled out I noticed flying was not such a lonely business as on ordinary nights. The city could be seen, lakes and rivers shone as London blazed fiercely; the ack-ack barrage was at 12,000ft. I cruised around, peering for some time before anything happened. When it came it was so sudden there was nothing I could do. An e/a flashed past so quickly and unexpectedly that it disappeared in the darkness before I could turn to follow. I decided to stay thereabouts; some more bombers would probably come along, I thought. Sure enough a second one appeared before five minutes were out. I kept him in sight as I dived and got under him so that he would stand out against the sky. Knowing how difficult it is to hit a swiftly moving target at night I held my fire until I could see the exhaust pipe flashes. It was an He111. Just as I was about to give Jerry a burst he suddenly dived. I followed him and gave him a few bursts on the way. He jinked sharp to starboard and banked, so I did the same, throwing confetti all the time. I got a dose from his gunner but no harm resulted. We were about 50yds apart with tracer trails criss-crossing between us. The Heinkel's fire improved as we levelled up; my aerial was shot away and he hit me more than once; luckily my engine remained untouched and ran splendidly all the time.

'The Jerry tried evasion action again – banks, dives, sudden pull-outs – he did everything possible to break away. By this time we were far south of London, but it looked as if the end was near; the German dived sharply with a long trail of smoke streaming from his starboard engine. My ammo ran out but I kept close behind him and had the satisfaction of seeing him splash into the sea. I don't know where it was exactly, somewhere between Dover and Southampton, about halfway. I got my bearing for base by the glare over London.'

At Redhill in Surrey the dawn brought to an end a night of lethal rich pickings for the Hurricane pilots of No. 1 Squadron. Many pilots from the cosmopolitan squadron were Czech, including Sergeants Bedrich Kratkoruky and Josef Dygryn. Kratkoruky argued with the intelligence officer that the unidentified bomber he had attacked over Canvey Island at 0030 was 'definitely destroyed'. But there was no proof, just Kratkoruky's insistence that when he had pulled out of his chase at 500ft the German 'was still diving towards the sea'. The intelligence officer apologized but said it would have to go down as a 'damaged' until it was corroborated by a AA crew or a Home Guard unit or another independent witness.

What 23-year-old Josef Dygryn from Prague had to report almost defied belief. Five minutes after Kratkoruky's contact over Canvey Island, he spotted a He111 'while circling anti-clockwise over London' at 17,000ft. The German was heading south, out of the guns of the Inner Artillery Zone towards home, cock-a-hoop at having survived the flak. Dygryn probed the Heinkel like a boxer sizing up the defences of his slower opponent with a series of jabs. He first made a 'quarter astern attack at close range', came in a second time, and a third when 'I closed right in and fired a long burst'. The alloy airframe of the Heinkel was famously robust but so was the Hurricane. It had a solid gun platform, its eight guns were grouped more tightly together than the Spitfire, and the wings gave it a rare stability. The other advantage over the Spitfire was that the Hurricane pilot had a perfect line of fire during an attack because the nose of his aircraft arced down slightly. In a Spitfire, a pilot sometimes lost sight of his quarry under the nose.

Dygryn's finger kept its distance from the firing button until the Heinkel was 20yds away. Then he pressed the tit. 'The enemy aircraft rocked violently and started to dive. I gave a final long burst on the dive and saw the e/a hit the ground.' Having emptied his eight Browning guns of 2,000 rounds, Dygryn returned to base at 0055 hours for replenishment. His ground crew bounded over to the Hurricane and prepared her for a rapid retake-off. The aircraft was reloaded and refuelled and within forty minutes Dygryn was airborne once more. He climbed north until he reached 15,000ft and almost at once his eyes lit up. 'I saw an He111 slightly above and heading South. I turned and climbed, giving chase until I was south of Gatwick.' Twice Dygryn opened fire but his prey wriggled free. The Czech manoeuvred himself into position for a beam attack. 'I gave another short burst and the e/a emitted black smoke which came over my cockpit. The e/a then burst into flames and went into a vertical spin out of control.'

Dygryn landed at Redhill at 2.05 a.m., half an hour after his second take-off. His ground crew whooped excitedly when he told them he'd shot down another German. The armourer re-armed the Hurricane, and applied a mixture of oil and paraffin to the guns to stop the dry cold of the freezing night disabling them. Oil, coolant and oxygen supplies were topped up. Shortly after 3 a.m. Dygryn was taxing down the runway on the hunt for a third victim. He flew east at 16,000ft, towards Kent and Biggin Hill, hardly daring to hope his luck would continue. Ten minutes of patrolling brought him a Junkers 88, 'at the same height as myself and in front, flying in a south-easterly direction'. Dygryn's left thumb moved over the firing button. He knew his shooting would have to be crisper with a Junkers than with a Heinkel. Junkers were wily opponents; their trick when chased was to extend their flaps and air brakes and virtually stop in the air. The fighter overshot and by the time he had turned round the Junkers was gone. But its weakness was its armament, with just a gunner in the ventral gondola to protect the rear.

Dygryn closed to within 25yds of the Junkers' tail and 'fired three 2-second bursts'. In fighter pilot's parlance, this was 'giving the Hun a squirt'. Dygryn watched cold-bloodedly as the Junkers began 'gradually losing height and emitting smoke. I followed him over the coast before I gave the final burst'. Flames from the engines wrapped themselves round the aircraft's fuselage as it shrieked down towards the sea. The gunner roasted in his gondola coffin. Dygryn made a rough note of where the Junkers had hurtled into the silvery waters of the Channel, '6 to 10 miles south of Hastings'. He landed at Redhill at 4 a.m. and made his report.

At dawn on 11 May the RAF were claiming to have downed thirty-three German aircraft. Many were unconfirmed and based solely on the pilots' accounts. They would be checked in time but the figure released to the press was thirty-three.*

At West Malling airfield Guy Gibson and Richard James had exhausted every possible profanity, such was their fury at their cannons jamming. Gibson rarely spoke to his ground crew, even if he'd shot an aircraft down; now they quailed before his fury. 'He could tear

*John Colville, Churchill's Private Secretary, wrote in his diary for 11 May that when he phoned the PM at Ditchley Park to tell him of the raid, the PM told him that 'we had shot down 45'.

people off a strip when he wanted to,' says James. 'Bully them, really.' Gibson wrote in his log book: 'Raid Patrol. A blitz on London. Saw two HE 111 – But Cannons would not fire. Damaged one with Brownings – no claims.' There was also, Gibson admitted later, 'a very rude remark written in the remarks column, but this was blacked out by the C.O.'.

Gibson got on his pushbike and cycled out along the quiet lanes of Kent to have breakfast with his wife. James retired to his billet in Malling Abbey and, as dawn broke over the abbey quadrangle, climbed into bed in one of the monastic cubicles.

26

AS Canon Alan Don walked back over Westminster Bridge a little after 6 a.m. he stopped for a moment and looked around him. 'The day seemed unable to dawn,' he told his diary, 'for a thick pall of smoke overhung London and kept the dawn at bay – it was as though the very Heavens were conspiring to add to the prevailing gloom.' The blackest, most malevolent cloud over Westminster rose from the shrunken ruins of the Electricity Supply Station in St Martin's Lane. It darkened the spirits of the Canon as he walked home, but at least St Margaret's 'had escaped unscathed'. As Don stepped over the myriad hose lines still patterning the grounds around Westminster Hall, he passed a group of bewildered policemen examining a short contorted length of railway line. In the clearing light they could see that Hungerford Bridge had been hit, but that was three-quarters of a mile down river. It wasn't possible, was it?

Don returned to his flat in Dean's Yard and kissed his wife 'good morning'. From the bedroom window he looked south and saw that the pinnacles of St John's Church in Smith Square 'were still burning with the dawn'. Charles Dickens would have been pleased to see the destruction of the 250-year-old church he'd called 'frightful and gigantic'.

The Stationmaster of Waterloo Station was also out and about when dawn came. The railways were in old Harry Greenfield's blood – his father had been Royal Train Guard to Queen Victoria – but he had never witnessed a night like the one just passed.

There had been sixteen major incidents at Waterloo and Greenfield had logged each one in his station diary as he sat in his office between platforms 15 and 16. High explosives, incendiaries, oil bombs and delayed action bombs, they'd had the lot. At first light the station was without gas, water, electric light and power; the signal box was dead, the lifts useless and much of the track severed. And as Greenfield began his walk along the track from Waterloo to Clapham Junction he became aware of a light rain falling, a strange rain. He held out his hands and caught one of the thousands of scraps of burned paper floating lazily away from what had once been the London Waste Paper Company warehouse in Belvedere Road.

Damage assessment was also taking place north of the river in another of the capital's main railway terminii, King's Cross. Two 500kg bombs had disembowelled the station at 1 a.m. but it wasn't until 3.30 a.m. that a parachute was spotted a little way up the track dangling from a signal. Driver George Irish and fireman Joe Cheetham had tip-toed towards the silk, half-expecting to find a dead airman. What they saw, lying between platforms 1 and 2 was the pillar box shape of a parachute mine. Bridling at the temerity of the Germans in having dropped a bomb on his workplace, Cheetham crept up to the 10ft bomb and placed his ear against it. He heard nothing. 'It's a dud, George.' Just before 6 a.m. the mine exploded.

Harry Beckingham had spent the night pursuing incendiaries in Pimlico and in the grounds of the Royal Chelsea Hospital. It wasn't until first light that his Bomb Disposal company, No. 35, was ordered to Victoria Station. Buckingham Palace Road was impassable to traffic so they walked to the station, wondering what they would find as they passed the Grosvenor House Hotel on the western side adorned with its many busts, among them Queen Victoria and Albert.

There were five unexploded bombs in total, recalls Harry. The entry point of the first he could see, 'just outside the Windsor Dive pub in the forecourt'. There was another UXB close by and three more inside. One bomb had landed on platform 2, another on platform 10 and a third had ploughed through a track just outside the station roof.

Harry's section walked into the station. The central clock on the main façade, set between two reclining sculptures, said 6.15 a.m. The whiff of flesh on the chilly morning air came from what was left of twelve horses from the Stag Brewery across the road in Palace Street, shredded by a direct hit an hour and a half earlier.

Harry was directed to the bomb on the tracks, already classified by an officer as category A; of utmost importance. The lonely walk took him within sight of the charcoal-black torso of a goods train on platform 10. Then he was there, standing over the hole of entry. Crouching on one knee, Harry measured the hole's diameter at a point a few inches below the surface. Any such hole, he knew, was normally about two inches larger than the bomb below that had created it. And a hole with a diameter 8 to 12ins across usually indicated a 50kg bomb, 12 to 18ins was 250kg and over 18ins meant they were facing a bomb 500kg or larger. Harry brought the tape measure up out of the hole. It was a 250kg bomb. 'The first thing we did then,' says Harry, 'was make arrangements for two railway wagons loaded with ballast to be shunted into place on the lines either side of the hole of entry as a blast wall.'

The most immediate challenge for Harry and his team was to expand the hole of entry. 'First we cut the sleepers on the tracks,' he remembers, 'and then we had to clear away the ballast from the edges.' With that done the soldiers started to chip away at the hole with pick axes and shovels. The path of a buried bomb varied from job to job for bomb disposal teams. Most hit the ground at an angle of 80° and continued straight for about 12ft, during which time the tailpiece and fins were wrenched off, before they jinked sideways or forward so that sometimes the bomb came to rest as much as 10ft away from a vertical line through the hole of entry. 'We found the fins early on with this one,' says Harry, 'which confirmed it was a 250kg bomb. What we wouldn't know until we found the bomb was the type of fuse, and that was what concerned us most.' The other slight worry in the back of their minds was what their exertions might do to the unseen bomb. 'Digging was always the hairiest part of the job,' says Harry, 'because the reason a bomb hadn't exploded was usually because it had been dropped too low from the aircraft to allow time for the charge to pass through the resistor into the firing capacitor. So it hit the ground inert instead of being fully charged. But there was still energy in the bomb and vibrations from people digging might well make it come alive.'

Victoria Station in the early morning, with bombs scattered everywhere, wasn't the place to entertain such thoughts. All they could do was keep digging. 'Too much thinking was unhealthy,' says Harry. 'I just kept to my philosophy that if a bomb went off and I was right on top of it then I wouldn't know anything about it. It would be a nice quick death.'

Digging was tiring work but by the time they saw the bomb's black body at least they no longer felt the cold. In a second or two they would know what sort of day it was going to be. 'Once we uncovered the bomb,' recalls Harry, 'we looked for the fuse number to see what we were up against.'

In the first month of the Blitz most UXBs had an impact fuse numbered 15. Bomb disposal squads soon mastered these by depressing the two spring-loaded plungers with a device called a Crabtree Discharger and removing the fuse. 'Then the Germans got more clever,' says Harry, 'and introduced the 17A fuse, which was a delayed action, liable to detonate up to eighty hours later. Worst of all was the anti-disturbance 50 fuse.' The 50 fuse was the latest and most lethal of German fuses. It needed movement of less than a millimetre to activate the trembler switch. Bearing such minutiae in mind Harry knew the bomb wasn't a 50 fuse, because if it had been he would have been dead long ago.

Harry scraped away at the earth around the bomb until he found the fuse pocket on the side. He read the number: it was a 25. 'That made life a lot easier than if it had been a 17 or a 50,' recalls Harry. 'The 25 was an impact fuse, but more advanced than the 15. If you tried using a Crabtree Discharger on a 25 you went up with the bomb, but it wasn't as tricky as a delayed action.'

The standard protocol in bomb disposal squads was for the men to do the digging and the officer to do the defusing. 'Usually though,' remembers Harry, 'after a heavy raid Lt Godsmark [their section officer] had so many other things on he would tell us to deal with some of them.'

The actual defusing of a bomb was a one-man job, so Harry was down in the hole by himself, alone with a 250kg cold steel bomb, 3ft 9ins in length and 14ins in diameter. Tucked away in his tunic somewhere was the pocket-sized book of regulations issued to all bomb disposal units. Rule two reminded men that: 'Bomb splinters may travel 1,200 yards. Therefore always keep under cover if possible. If you must be in the open, lie down. When lying down, men are reasonably safe against a buried bomb at 100 yards and against a bomb on the surface at 300 yards.'

Harry placed a BD Discharger against the bomb. 'This was a container with a bicycle valve at one end and a tap at the other,' he says. 'Inside was a solution of salt, benzol and methylated spirits.' Harry attached the tap over the bomb's fusehead. 'Then I started pumping the solution into the fuse and it was passed through the plungers under pressure.' Two hundred yards away the rest of Harry's squad listened to the pumping. When all the solution had been forced into the fuse, he climbed out of the hole and walked back along the tracks to the men 'waiting with a cup of tea at the safety point'. Harry sat and supped his tea as the new day began, glancing from time to time at his wristwatch. Having waited thirty minutes for the solution to short circuit the fuse, Harry left his pals and walked back down the tracks making patterns in the air with his breath as he slipped between the two railway wagons and eased himself into the hole. 'I checked the container to make sure all the solution had been pumped into the bomb,' recalls Harry. 'It had gone in so the bomb was almost definitely inert. At least, you hoped to God it was inert. Then I took my fuse key, locked it onto the fuse ring and twisted. Once the ring was unscrewed I took out the fuse and removed the gaine so it was completely safe.' Harry slipped the gaine in his pocket and 'climbed the hell out of the hole'.

AT 7 a.m. the Elephant & Castle was drinking water from 9 miles of hose. As the Thames rose so it became easier to pump and the fires that had run riot for seven hours were finally curbed. Men, women and children emerged from the bowels of the earth into a charnel house. The yardmaster of the Elephant & Castle station remembered the 'look of amazement on their faces as they made their way out of the station and saw the amount of destruction that had taken place during the night whilst they had been down below safe and sound'. The yardmaster had stashed the station's money safely but none the less he was reassured by the presence of the Home Guard, 'about in the street on account of some looting that was started by some of the rough element in the neighbourhood'.

Looting was a profitable canker in the streets that encircled the Elephant. A posse of firewatchers, run by a 'gorilla-like ex-wrestler', patrolled up and down the Walworth Road during the Blitz, their steel helmets and armbands worn with brio. When there was no one around, one of them would heave a brick through a shop window and then his pal stepped through the broken glass and removed the goods. The Reverend J. G. Markham, rector of St Peter's, and a warden too, knew what went on but 'I never reported anybody to the police, who during air raids, except for the occasional war reserve members of the force, were conspicuously absent from the scene'.

The moment the siren had sounded at 5.52 a.m. the looters went to work, slithering unseen into damaged houses and wrecked flats and helping themselves to the spoils of war; gas and electricity meters were rifled expertly and any mother foolish enough to have displayed her son's First World War medals was punished for her pride. Often the looters sent in their children; on one Friday the previous April, Lambeth Juvenile Court had a sorry procession of forty-two offenders to deal with. Many were teenage girls who wandered over bomb sites taking clothes from the dead. But the youngest defendant on this particular Friday was a 7-year-old boy who had stolen five shillings from a gas meter.

'There was a lot of looting going on,' Emily Macfarlane remembers. 'People came down here and just helped themselves. A jeweller's in the Elephant & Castle had been hit and there were watches and bits of jewellery lying all over the place. People just picked them up and filled their pockets.'

Joan Veazey was revolted by what she saw going on around the Elephant & Castle. 'The most sickening thing was to see people like vultures, picking up things and taking them away. I didn't like to feel that English people would do this, but they did.'

Sixteen-year-old Joe Richardson remembers being 'not that shocked' when he poked his bleary-eyed head out from the shelter in John Ruskin Street. 'I think I'd got used to it all by now,' he says. Joe went back to check on his aunt in Bowyer Street and, with the curiosity of his age, began to explore his neighbourhood. He walked up the Walworth Road, 'with hoses everywhere', to the Elephant & Castle. Now he was shocked. The flames had been conquered but what destruction they had left behind. Joe heard a whimpering from behind. He turned and 'there was this big dog hobbling towards me. His fur was all burned off and his skin was sort of bubbling'. A Home Guard soldier raised his rifle and shot the dog.

THE machines were running in the *News of the World*'s offices when the All Clear sounded. 'We printed in a red-hot building,' recalled Ernest Riley, 'but the *News of the World* went out as usual.' The lead story was about the struggle to oust the pro-German politician Rashid Ali from power in Irak [sic]. Underneath the headline was literally hot off the press: 'Big Raid This Morning: Bomb-a-minute Blitz.'

'Nazi raiders came over London in force shortly before midnight – the night of the full moon – and early this morning the attack developed on a big scale.'

At the *Mirror*'s offices in Geraldine House Cecil Harmsworth King was simmering under his firewatcher's hat. They were trying to get the *Daily Mirror*'s Sunday sister printed, *The Sunday Pictorial*, but the electrics had shut down. 'However, after a ten-minute interval they came on again and efforts could be made to get the paper out.'

King picked up the first editions and admired the bullish headline: 'Many Nazi Raiders Down: Great London Blitz'. The copy underneath was laced with similar swagger. 'Nazi Bombers blasted London in brilliant moonlight last night – and they paid a heavy price. Soon after the alert – it was before midnight – roof watchers saw the first raiders fall in a whirl of flame on the outskirts of London.'

The *Pictorial* made a fleeting reference to the damage inflicted by the Luftwaffe towards the end of the article. 'Among the buildings hit were two hospitals, a warden's post, an AFS post and a club ... two famous London churches were set on fire.' There was also a report on the Wembley Cup Final, in which the paper's sports correspondent had used an unfortunate turn of phrase: 'It was Blitz football!' he shrilled. 'Thrill upon thrill.'

Jim Goldsmith, one of the firemen still fighting the fires in and around St Paul's Cathedral, recalled 'returning to the station for more hose, fed up, dirty, soaking wet and still trying to stop the fires spreading'. Outside the station was a pile of one Sunday newspaper delivered for the firemen. Goldsmith remembered one of the firemen picking up a paper and laughing. His mates looked over his shoulder and read the headline: 'A RAID TOOK PLACE ON LONDON TONIGHT BUT BY THIS MORNING THE FLAMES WERE UNDER CONTROL'. 'How is that for faith?' wondered Goldsmith.

For the reporters of the daily newspapers the All Clear signalled the start of the race for the biggest scoop. John Hughes was asleep in the Reuters dormitory when he began 'to hear somebody saying a long, a very long, way off "Wake up! Wake up! The Abbey's on fire!"' Hughes ran up to his office on the fifth floor and was at once grabbed by Cole, the night editor. 'Slip down to the [Westminster] Abbey, old boy, and get a story through as quickly as you can, America is waiting for it and the Dominions.'

Hughes stuffed his notebook in his pocket, slipped his pencil inside his jacket, and careered out of the Reuters building. It was just before 7.30 a.m. as he headed west up Fleet Street. 'I hardly met a soul,' he told his daughter in a letter, 'except long convoys of AFS mobile water units.' He encountered one other journalist, Paul Mitchell, a fellow Australian working for the *Sydney Morning Herald*. 'He said "you ought to see Aldywych and Catherine Street just near the Waldorf [Hotel]".' Hughes hurried on to Catherine Street where he stared in disbelief at a 'hole about 50ft wide and 20ft deep'. 'Opposite Australia House a big building, Accident & General, was on fire and all the roadway was impassable with huge blocks of masonry. On the other side at the corner of Norfolk Street* and the Strand, a big building had been hit by an HE and had fallen into the road. That horrible death mask (as I always call it) of a big bomb – a pall of horrible dirty-looking dust – hung over everything.'

Hughes continued west down the Strand towards Trafalgar Square. As he passed the front of the Savoy Hotel, Larry Rue of the *Chicago Daily Tribune,* Jamie Macdonald of the *New York Times* and Quentin Reynolds from *Colliers Weekly Magazine* were about to join the search for a scoop. They chose to go east, the way Hughes had just come, three Americans and an Australian united in their incredulity at the mayhem that confronted them. In his report for *Colliers*, Reynolds told America how a small group of people 'stood staring at the smoking wreckage of St Clement Danes which was Doctor Johnson's church; every child in London has sung the nursery rhyme which begins "Oranges and Lemons say the Bells of St Clements". The church was still burning and the light from the flames within showed softly through the stained glass windows'. Reynolds noticed that the statue of Dr Johnson at the rear of the church was undamaged. Thank Heavens the doctor faced east towards Fleet Street, he thought, so at least he was unable to see the sorrowful sight behind him. In Fleet Street Reynolds was alarmed to see the damage inflicted upon the offices of the United Press agency in Bouverie Street. But his dinner partner of a few hours earlier, Ed Beattie, was unharmed. Unharmed, but disappointed that he wouldn't be able to take Reynolds for the bicycle tour of Kent he'd promised him for later that morning.

Larry Rue described in his Isolationist newspaper how he 'took a walk amid broken glass, heaps of debris and acrid smoke, stepping gingerly over fire hoses'. With his pencil he painted pictures of rescuers

* Norfolk Sreet ran south from the eastern end of the Aldwych to the Embankment. It no longer exists.

digging out the dead, of ambulances waiting for the wounded and of carnage everywhere. 'Amid the changing scenes of the morning's walk,' he told his readers, 'one stands out, although irrelevantly. When the rising sun was obscured, behind sulphurous clouds of smoke, pigeons and sea gulls stood side by side, huddled compactly together, on the rail of a bridge. They faced the sun but did not move, seemingly bewildered whether this really was daylight or not.'

Macdonald turned to Londoners to flesh out his copy. 'One firemen, asked what he thought the raid amounted to, said "Mate, this was the worst ever! You should've been here and seen it and 'eard it! Blimey, what a night!"' As he tramped the streets of London Macdonald marvelled at the 'innumerable splotches of sand where incendiary bombs had been snuffed out'.

Hughes meanwhile had stolen a march on his American rivals. He was now coming within sight of the Palace of Westminster and even though 'the air was acrid with the smoke of the 2,000 fires', he could smell a scoop. 'As I approached the Abbey I thought Big Ben looked a bit wonky. I studied the great tower where the world famous clock strikes the hours ... it was just on eight o'clock. As the hour hand crept on to the figure eight there wasn't a sound. Big Ben, for the first time since the war, had been reduced to silence by the Hun at last!'

At the pillbox disguised as a W. H. Smith bookstall in Parliament Square, Hughes fell into conversation with the 'steel helmeted fully armed sentries there' and gleaned the gist of a 'nice little story for the PA [Press Association]'. He sprinted to a public phone and picked up the receiver. The phone lines were working once more. 'I phoned through to the PA what we call a "snap" of Big Ben. That is, just the fact that the famous clock had been silenced and damaged by German bombs ... I then went along to the Abbey for the biggest story of my life.'

Canon Alan Don and John Colville, Churchill's Private Secretary, were also at the Abbey at 8 a.m. for Holy Communion, a weekly celebration since 1847. 'But it was found impossible to hold a service,' said Don, 'and of course the other services were cancelled.' Don left the Abbey and went to view the destruction in the House of Commons, where he learned to his deep regret that Captain Elliott 'was actually struck by a bomb as it fell besides him ... and two auxiliary policemen were killed on the roof'. Colville was turned away from the Abbey by a policeman at the door. 'There will not be any services in the Abbey today, sir'. Colville 'stood on Westminster Bridge and thought ironically of Wordsworth and 1802. St Thomas's Hospital was ablaze, the livid colour of the sky extended from Lambeth to St Paul's, flames were visible all along the Embankment, there was smoke rising thickly as far as the eye could see. After no previous raid has London looked so wounded the next day'. Colville returned home, breakfasted, 'then rang up the PM at Ditchley and described what I had seen. He was very grieved that William Rufus's roof at Westminster Hall should have gone'.

From the House of Commons Don had returned home to 20 Dean's Yard and discovered a short time later that his faith, or his luck, had saved his house. 'Incendiaries had penetrated the roof of the attic, set light to a bit of the passage floor and for some inexplicable reason, burnt itself out without starting a fire.'

On the Abbey's flagstones Hughes's footsteps provided a mournful echo as he gaped at the desecration all around him. Up among the black-scorched hammer-beams a pair of Black Redstarts chirped as they built their nest. Hughes later described to his daughter how: 'I saw sitting in the darkened Lantern, the debris of which lay scattered on the floor of the nave as I approached him, the most Rev Dr Jocelyn Perkins ... Sacristan of the Abbey. The old boy didn't seem particularly perturbed. In fact he told me that the damage looked far worse than it really was and that the fabric of the Abbey had not been touched at all.'

The 71-year-old Perkins was bald and as deaf as a post, so Hughes let him do the talking. And he listened well, because no one on earth knew the Abbey better than the arthritic Perkins. He had been appointed to the dual position of Sacrist and Minor Canon on 12 December 1899 and arrived the following February on the day that the bells were rung to celebrate the relief of Ladysmith in the Boer War. Perkins was a traditionalist with a reputation for smiling fierceness – in 1941 he was still chuntering away to himself about the electrification of the Abbey in 1911 – and any miscreant who irritated him would be admonished with the familiar refrain, 'Now come along – don't be an owl'. But now he talked to Hughes with the breathless delirium of one glad to be alive after his night spent in the concrete shelter in College Garden. Perkins's dark spraying eyebrows wriggled with excitement as he enlightened the Australian journalist 'about the Lantern Tower and he took me in front of the High Altar where the kings and queens of England have been crowned for hundreds of years'. Hughes checked his historical and architectural facts with Perkins and then asked to see some further damage. 'The old boy took me through the cloisters into the College Garden – the oldest and loveliest garden in Westminster, a dream of loveliness in normal times in May [now] covered with filth and debris of every description.' Perkins showed Hughes the seventeenth-century houses in Little Cloisters. His house, No. 5, where he and his wife, Emma, had lived for forty years, was intact, as was No. 4; Nos. 1, 2, 3, 6, 7, including those occupied by Canon Michael Barry, Dr Ernest Bullock, the Abbey organist, and the Precentor, Rev. C. M. Armitage, were sepulchral relics. 'Lovely bits of Sheraton and Chippendale lay piled up on the lawns, with carpets blackened with oil from oil bombs, bits of smashed vases, all the precious things that people collect over long years and cherish and value.'

An impromptu gust of wind stirred the embers of Dr Bullock's charred house and a host of vengeful red sparks took off towards Perkins's house. He rushed over to harangue some firemen resting after dampening down the fires in Westminster School. They looked wearily at the Sacristan, then at his unscathed house, and they told him to relax.

Hughes left the fretful Perkins and walked towards Westminster School, right into the path of A. L. N. Russell. 'The school had been irreparably damaged,' Hughes said, 'and here was its architect to tell me all about it. He was in Home Guard's uniform and was on duty. Mr Russell had certainly seen something that certain night.' Russell steered Hughes through the school where in one of the form rooms they stood on a floor that was a patchwork of roof timbers, slates and metalwork. A few books and photos had survived and Russell pointed out one in particular to Hughes. 'It was of an athletic team in 1937 and the plump and confident-looking boy at the end of the row was the young Ribbentrop, whose entrance at Westminster was arranged as a special courtesy though he was past the usual age of admission.'

After the school Hughes's tour moved to Westminster Abbey. As he and Russell stood under the Lantern roof in Westminster 'discussing architectural details so that my story would be right technically as well as journalistically ... there was a fall of large pieces of bricks and things. We just had time to step aside when the place we had been standing on was buried beneath several cwt of debris'.

Hughes was sure he'd won the race for the best scoop. 'What a night, what a sight, what a story! ... I got a taxi back to the office and dictated it all in about half an hour. I think it was a good story.' Hughes included the near miss under the Lantern roof. At 8.30 a.m. he handed his piece to the editor-in-chief, who sat down to read.

> Westminster Abbey stands open to the sky where the roof over the lantern in the centre of the building has fallen in, but the Dean,

Dr. P. F. de Labillière, said that the main fabric was intact, and no irreparable damage had been done.

The Deanery, one of the most perfect medieval houses in England, has been destroyed, and the Dean and his wife have nothing left but the clothes they were wearing.

Part of the wreckage of the Dean's house has fallen on Cloister Garth, the square of turf in the middle of the cloisters. The cloister walks are flooded, but the cloisters are unharmed.

The roof of the lantern was destroyed by an incendiary bomb. The roof timbers were all burnt, and the timbers and vaulting fell to the Abbey floor, where the transept and the choir intersect.

While a reporter was inspecting the wreckage with Mr. A. L. N. Russell, architect to Westminster School, several tons of masonry from the injured lantern fell within a few yards of them. The pulpit has been partly destroyed. A considerable amount of damage, Mr. Russell said, must be done by the water which is streaming down from the roof, but the outstanding fact is that the most historic parts of the Abbey are uninjured.

The eastern part of the Abbey, where the royal tombs are situated, is intact. 'The fact that a 19th-century roof has been destroyed is not an irreparable thing, although the Abbey is now open to the sky,' said Mr. Russell.

The top side of one of the pinnacles of the Western side of the south transept has been slightly damaged. In addition to the Deanery three houses in Little Cloisters were burned out. Canon Barry's house, the house of Dr. Bullock, the organist, and No. 2, the house of the precentor, the Rev. C. M. Armitage.

Dr. Perkins, the Sacrist, gave a reporter a vivid account of the scenes in the Abbey when the roof of the lantern fell in. 'In spite of every effort by a large number of firemen and our own firewatchers,' he said, 'we were unable to get the flames in hand before the incendiaries destroyed the roof of the lantern, the pulpit, and a great many of the pews. Before this happened we had to endure the agonising sight of the lovely houses in the college gardens occupied by Canon Barry and Dr. Bullock going up in flames.

'Then the Deanery went. Dr. de Labillière and his wife inspired us all by the calmness and fortitude they displayed in the face of the loss of their lovely home and of every stick of their personal belongings.

'They stood on the lawn, with the fires burning all around them, concerned only for the safety of others and the efforts of the firemen to save the Abbey. Only after he had done all he could did the Dean consent to snatch an hour or two of rest in my house, but he was up again at dawn, consumed with anxiety for the fate of the Abbey. His relief when there was sufficient light to reassure himself that the damage, though serious, had not affected the main parts of the building was touching to see.

'But for the A.F.S. men and our own fire-fighters, who put everything they had into the fight to save it, the Abbey must have been destroyed. The firemen succeeded in preventing the flames from spreading, and we are thankful to God.'

In the House of Commons, the debating chamber has been wrecked and it is feared that it cannot be used again until it has been rebuilt. Bombs have also smashed the roof of the Members' Lobby, which had already been hit in previous raids and was shored up by elaborate scaffolding. Doors were torn off and windows smashed.

Fire in the Houses of Parliament began along the side near Cromwell's statue, which was damaged in an earlier raid. Victoria Tower and Westminster Hall were later involved, and the main fire caught the Chamber.

At one time it was feared that Big Ben had been destroyed. A great crash of tumbling masonry and clouds of dust and mortar gave

> that impression to fire-fighters, but they saw when the wreckage had settled that they were wrong. The clock faces were blackened and scarred, but the clock still tells the hours.
>
> It is understood that in the House of Lords, Captain E. L. H. Elliott, resident superintendent, was killed at his post while helping with the fire-fighting, and that others killed were two members of the police war reserve and one custodian. Other members of staff who worked right through the night putting out the flames and salvaging valuable records and relics had remarkable escapes when the bombs fell.
>
> What some consider the most magnificent roof in the world, that of Westminster Hall, with its soaring arches and sweeping beams of oak, was pierced by bombs, and damage was done to the interior ... The Upper School is a charred and roofless wreck. The four walls scarred and burnt, remain standing ... and on each of them can still be traced only a very few of the multitude of names that were inscribed on them. The lower floor of the college, made up of the king's scholars' living-rooms, has become the resting place of rubble, burnt timbers and twisted girders.

'The editor-in-chief congratulated me,' Hughes wrote to his daughter, 'and told me it was a "fine piece of work" and all my colleagues here were charming.' The story was sent to the censor, and Hughes at last put up his feet.

ACROSS London others hadn't the luxury of a break. Reenie Carter at 'A' Division HQ in Dean's Yard was still logging calls. A fire in Victoria Street, running east from the station, was reported just after 7.30 a.m. By 8.30 a.m. Reenie was faced with a huge pile of incident reports. 'It didn't matter that we'd been up all night in an air raid, we still had to log every single incident in chronological order in the book, and if we made a mistake we had to get the commanding officer to sign it. It took us hours to write everything in the log book.'

The flames hadn't finished with St Mary's Church in Newington on Sunday morning either. They continued to spring to life, intermittently and desultorily, though there was little left to burn. Christopher Veazey held Communion in the church hall at 9 a.m. and Mrs FitzGeorge, wife of the rector, saw no reason to cancel the Mother's Union scheduled for later in the morning.

In Wormwood Scrubs Park, Bombardier Bill Church and the rest of the crew had just finished changing the barrel of their 3.7in AA gun. 'Most barrels wore out after about 800 rounds,' says Bill, 'because you've fired so many rounds that the shell's copper driving band doesn't engage in the rifling and ultimately the muzzle velocity drops. A good gun crew could change a barrel in about 35 minutes but you never did it during a raid.' With that done the men crawled into their bivouacs and were asleep in seconds, oblivious to the cold, oblivious to their pounding heads, oblivious to the shell cases piled high around them. They would be sent back later to ordnance for recycling.

In the vaults under Waterloo Station the fires near the Lower Road continued on their merry way. Firemen clung to their hoses, retching from the searing heat and sneezing from the rising dust. From deep within one vault there was an outbreak of what sounded like machine-gun fire. Firemen threw themselves to the ground and looked to each other for an explanation. They found one a little later, when they came across scores of champagne corks from one of the bonded warehouses now razed beyond recognition. With the mains outside the station dry, getting water down to fight the seat of the blaze continued to tax the firemen. A supply was found in the catch pits of the Waterloo and City Railway's berthing sidings, but how could they get a pump down there? It was finessed through the Waterloo and City workshops, lowered by a crane and then pushed over the rails to the catch pits. From there the water was pumped back over a 30ft wall and relayed to the fire in the Lower Road. Slowly, with a grudging reluctance, the flames retreated, though they made a fighting withdrawal from Lower Road on the south and York Road on the north. Sporadically they found a combustible material on which to feed and as the firemen advanced more confidently, an invisible dragon seemed to roar out a plume of vicious fire.

At the Temple, remembered J. W. Morris, the barrister who had striven manfully to save the church, 'it must have been about 9 o'clock on the Sunday morning before ... firemen found hydrants some distance away' to tackle the furnace that had run amok among much of the Inner and Middle Temple. After an arduous journey through the night from Oxford a crew arrived, guided every step of the way by the flames over London. They were too late to save the church or the Inner Temple library or Hall. Lamb Building too was beyond redemption. Residents attempted to rescue what they could from the offices still threatened by the flames. In the Treasury Office the sub-treasurer, Roy Robinson, and his wife removed fittings and furniture. 'I found my husband struggling to remove a picture from the wall,' recalled Mrs Robinson. 'I said "Well, he's dead anyway and it is no use killing yourself to save him". In spite of this I think the picture was saved.'

In Queen Victoria Street the battle to save Faraday House still hung in the balance at 9.30 a.m. A little over three hours earlier Major Frank Jackson, chief officer of the London Fire Service, had cancelled all officers' leave and mobilized 1,000 firemen from outside London. More canteen vans flooded into the city now, dishing out food and hot drinks to the dazed firemen. Reinvigorated, the firemen renewed their efforts, but the fires wanted Faraday House. On one of the fireboats anchored in the Thames below Queen Victoria Street they were pumping water up from the river and feeding it to the firemen on the shore through their hose. That in itself had been a bitter struggle, a duel against the dark clotted mud revealed at low tide on the Thames. One of the firemen, Dick Helyer, remembered that 'you had to get your hose to the water's edge, try to get it over the mud and have a line to give it to the chaps on the shore to pull it ashore, but the mud used to come up to our knees and you had a hell of a job'.

Then the tide began to push Heyler's fireboat down river towards Southwark Bridge. They arrested their drift and resumed pumping; what they were pumping now, recalled Heyler, 'looked as though it was boiling … the fires were consuming so much of the oxygen in the atmosphere that you got a down draught which went along the top of the water into the base of the fire'. On another fireboat by the downstream side of Blackfriars Bridge a fireman spotted something else bobbing up and down on the tide. 'Scorched brown and horrible with burnt hair and what looked like elbows … what got me most was that this thing had no head.'

27

AT 10 a.m. the sun over London glowed blood red through a sky as yellow as one of the city's choking winter smogs. Ash lay like a freshly fallen snow on the ground. Trees and hedges were festooned with articles of clothing blown out of homes. Quentin Reynolds had finished surveying the ruins; now he watched deadpan Londoners sifting through the detritus of their city. Gone, he noticed, was the complacency so evident during the autumn of 1940, the 'We Can Take It, Jerry' bravado that had captured the imagination of the American journalists. 'I walked around the still burning streets of London on Sunday morning,' he wrote to his readers. 'The streets were filled with grim-faced, sullen-looking men and women. They were through "Taking It". They wanted to give it. Every bomb that the Nazis dropped during the night carried germs with it – germs of hatred. I could feel the hatred rising from the ruins infecting everyone. Tight-lipped men and women stared at the debris of treasured landmarks and you could feel the hatred of Nazi barbarism emanating from them.'

They stared at the church of St Mary-le-Bow in Cheapside, 'Bow Church' to Cockneys, those people born within the sound of its twelve Bow Bells*. The BBC had been using the peal of the bells to signal

* This legend dates back to the fourteenth century and the ringing of the bells at 9 p.m. to signal the city's curfew.

the start of each wireless broadcast to Occupied Europe. A church had stood on this spot since 1087; it had been rebuilt by Christopher Wren six hundred years later after the Great Fire of London at a cost of £15,400, his most expensive church. Now all that remained was the steeple with its dragon weathercock, two outer walls and the oak pulpit with its monogram 'CC', representing King Charles II and Queen Catherine of Braganza.

They stared at St James's Palace, built by King Henry VIII in 1531 and proudly redolent of England's glorious past. From St James's Elizabeth I had set out to address her sailors at Tilbury on the eve of the Spanish Armada; Charles II was born in the Palace, so too James II and Queen Anne; Queen Victoria married Prince Albert in the Chapel Royal in 1840. A stick of high explosives had destroyed much of Friary Court, the walls were pockmarked and scarred and the famous clock that faced up St James's Street had been smashed.

They stared at the Old Bailey, London's Central Criminal Court, with every one of its windows smashed and the north-west corner, including Court Two, flattened by a huge bomb. The gilded Lady of Justice was unblemished but the floor of the lavish marble hall was coated with masonry. Not all Londoners lamented the wounding of the Old Bailey; one picked his way through the mess into a room on the Sheriff's Corridor and removed the judiciary's supply of 1,000 finest Havana cigars.

They stared at the British Museum, and the museum's director Sir John Forsdyke, as he sat in the quadrant beaten and broken, overwhelmed at the list of galleries destroyed: the Roman Britain Room, the Greek Bronze Room, the Coins and Medals Room, the Central Saloon and the adjoining Prehistoric Room, the Fourth Vase Room and the Greek and Roman Life Room. Some of the exhibits had been moved but many were gone; the roof of the main staircase was no more and the incendiaries that fell on the south-west quarter of the museum with an innocuous plop-plop had feasted royally on the bookstacks. A soft breeze now blew the black skeletons of 150,000 books out over the museum's quadrant.

They stared at the Queen's Hall, bereft of its interior and looking like the remnant of a Roman arena with just the outer shell left. Musicians from the London Philharmonic Orchestra arrived at the stage door early in the morning for a 10 a.m. rehearsal of the 'Sunday Concert' scheduled for the afternoon. 'There'll be no rehearsal here this morning,' the hall-keeper told them. One of the musicians, a horn player, asked about the instruments. 'There'll be ashes by now,' replied the hall-keeper. A cellist said he would see for himself and moved towards the door. 'You can't,' the hall-keeper told him. 'It's a blazing red-hot furnace at the end of the passage.' Instead the cellist gave a fiddler a leg up through a window. He disappeared inside and returned a few minutes later with 'a scorched leather case'. It was claimed by a horn player, who opened the case and took out his French Horn. 'I'll fix it up and see if it works,' he said. 'I bet it's as good as it ever was.' He played a few notes from Beethoven's Fifth Symphony, 1st Movement. 'Never say die!' he grinned at his colleagues. The rest of the orchestra foraged through the litter of their instruments. Cedric Sharp took away the corpse of his £600 cello, determined to restore it, but the magnificent Italian double-bass of Adolf Lotter, the London Symphony Orchestra's principal, was a sickening chunk of charred wood and strings. One of them took a sign from the caretaker's office and hung it on one of the entrance doors: 'All seats sold out'.

Westminster Palace was used to the stares of sightseers, but the people now gathered outside were Londoners who saw through the smoke the smouldering House of Commons. The anger they felt was the same as that among the crowds at St Clement Danes, Lambeth Palace and the Temple.

Quentin Reynolds finished his walk at the Houses of Parliament.

He stood before the statue of Abraham Lincoln, 'with the House of Commons to his left and Westminster Abbey to his right'. Reynolds scrutinized Lincoln's face: 'He looked grim but understanding. From the Nazi point of view, both were military objectives and Lincoln, brooding there solemnly, looked as though he understood.' And Reynolds knew that the English understood what had to be done; no more chivalry, no more 'fair play'. 'All that day I sensed a new and intensified hatred of Germany in the people of London,' he reflected. 'The steel had entered the soul of Britain.' Ambulance driver Anne Spooner says it was at this point 'I started to hate the Germans'. A young woman, Moyra Macleod, who arrived in London after the raid to attend an English Speaking Union lunch looked around her and wrote in her diary: 'I hate Germans and hope we blow the whole bastard lot of them to pieces with their Berlin around their ears.' Reg Harpur was also consumed with hatred. He had watched the raid unfold from his doorstep in Sydenham and returning firemen told him about the damage to Parliament and Westminster Abbey. 'My hate of the Germans has increased, if that was possible ... I hope that we decide to give their most valued buildings the same treatment.'

Gladys Shaw's Sunday School class in Peckham was smaller than usual this morning, though the message was the same. Love thy enemy. 'I tried to make the children know that it was war that was wrong and that the airmen who had done the bombing had mothers too, just like they did.' She asked the children to join with her in saying a prayer for the Germans. 'Afterwards one little boy told me that he had kept his eyes open when we said the prayer. He didn't want to pray for them.' After the school Gladys walked home to her bomb damaged house in St Mary's Road, past row after row of stricken homes. 'The parish looked a broken jaw with its teeth hanging out,' she says.

At her husband's army camp in the south-west of England, Joan Wallstab had made a rare foray into church. 'I had never been one for organised religion,' she says, 'but Eddie had to go with his regiment so I went too as it meant I could be with him.' No one in their peaceful corner of the country had an inkling of the fury that had passed over London hours earlier. Joan was sitting next to her husband when 'something came over me and I burst into tears'. Her husband led her out of the church and soothed her fears. 'I couldn't explain why I had started crying but I was scared.' An hour or two later news of the attack reached the camp. Joan picked up the phone and asked the operator to connect her to her parents' number. 'The phone rang and rang but no one answered.' She tried again later. What she would have given to hear the steady Norfolk burr of her father's voice at the end of the phone, laughing at her anxiety. But no one answered the phone, not even the rescue workers searching for survivors in the rubble of Newington Causeway's Sessions House. 'When I couldn't get any answer,' recalls Joan, 'I knew something had happened so Eddie got permission to come to London with me.'

As Joan started out for London, plump Bob Post of the *New York Times* was at his improvised bureau in the Savoy, filing his copy. He articulated with moving eloquence the clawing, flailing suspense that some Blitz survivors felt after every raid. 'It is a curious feeling that overtakes people the morning after a bad Blitz. The first thing you do is pick up the telephone to try to talk to your best friends. All day you telephone to ask people about other people. You say "Glad to see you"and the phrase has a special meaning. Every familiar face is chalked up as a sort of special personal victory as if one's own good fortune in surviving by careful precautions were reinforced by the fact that one's friends had had the same experience.'

Along the Peckham Road John Fowler had woken on Sunday morning with a throbbing head. Too young to have the worries of others on his mind, all he knew was that he was alive and it felt good. There was no gas, no water, no light, the house was freezing but he was alive. Suddenly he was conscious of how banal life was in the countryside; out of harm's way, the tedium broken only by the worry of life's petty problems. Life in London was lived with such powerful intensity. It was great, so long as your luck held. He left his house in East Surrey Grove with his parents, all of them determined they would reach Sturminster Newton for his sister's thirteenth birthday. 'Our road had hardly been touched,' recalls John, 'the odd tile down and a few windows broken but that was it. Going up to Waterloo was a different story. There were no buses obviously so we walked.' They headed west on to the Walworth Road and skirted round the Elephant & Castle. 'It was just one big mess,' remembers John. 'The roads were blocked, there were hoses everywhere, rescue teams digging people out, the dead lying all around. Sometimes you saw bits of the dead.' John and his parents walked down Waterloo Road, 'past The Cut which had taken a pounding and they were still putting out fires at Waterloo Station'. The Fowlers joined the scores of people milling around outside the entrance. 'It was all very calm and orderly,' he says. 'They told us there were no trains from Waterloo but that buses were on their way to drive us to Clapham Junction, and would we mind being patient because they might take a while.'

They stood waiting in line for a bus a little distance away from Waterloo, 'and then the strangest thing happened', says John. 'A girl appeared on the other side of the road, looking straight at me. She smiled, waved at me and I'm sure she shouted "Hello John". Then she disappeared in the crowd.' John waved back belatedly. 'Who was that?' his parents asked. 'It was Rose,' replied John. 'I could've sworn it was my cousin, Rose. But as I found out later, it couldn't have been, because on Sunday morning she was buried under the rubble.'

28

JOHN'S cousin was one of hundreds unaccounted for on Sunday morning. Dead, wounded or alive and well, no one knew. In theory ARP wardens kept census sheets with details of who lived in which street or in which block of flats, and where they went during an air raid. In practice, it wasn't that simple. People changed their plans. Someone who always stayed in his own house might decide on a whim to go to the shelter, without telling a warden. Another person who sought sanctuary in the shelter might stop over at a friend's. At the Columbia Buildings in Hackney rescue squads had already brought out six people. They asked the local warden if that was everyone. The warden scanned his list. Alice Desert and her husband George from No. 52 hadn't been seen, but they always went to the shelter. The warden gave the All Clear, unaware that the Deserts had invited round their 21-year-old daughter and son-in-law the previous evening. Children were soon playing in the sunshine amid the wreckage and it wasn't until Monday that a quavering cry was heard from the rubble. Only Alice Desert was brought out alive but she died of her injuries two days' later. 'It seems possible to me that her life may have been saved if she had been released on the Sunday,' her nephew wrote to the local Civil Defence Authority in his search for answers. 'I hardly like to think what her feelings must have been when she heard children playing on the debris whilst she was trapped and her husband was lying at her side dead. If wardens were not aware that people were still buried, then I suggest that they are guilty of serious neglect of their duty.'

In Walworth Road rescuers found Austin Osman Spare lying among the frames of his canvases, now cheap splinters of black wood. His half a dozen cats were on the rubble holding a vigil. He was pulled from the wreckage and laid on a stretcher. His beard was singed and the glint in his eye was wilder than ever. He told nurses that he had no feeling down his right side, not even in his arm, his painting arm.

In Vestry Road in Peckham a rescue squad was searching for John Fowler's cousin Rose. They had found the body of her dad but there was no trace of the 15-year-old daughter. Across London the same efficient routine was being repeated by rescuers. A call for silence from the leader of the squad; a yell into the wreckage. 'Can you hear me?'; mounting tension as they wait for a cry or a call. Tap, tap, tap. What's that? Someone signalling to them from down below, or just debris gently shifting its position.

The men of the rescue squads were famous as much for their impudence as for their bravery. They weren't averse to looting while they rescued – six members of one south London rescue ream had recently been imprisoned for nine months – and they often spirited away equipment from other Civil Defence organizations. But stories of their courage were legion; most rescue teams were composed of pre-war builders and labourers, hard men who lived by a different set of rules to the journalists and politicians who rushed to condemn them if they erred.

Builders made the best rescuers, reflected Kenneth Sinclair-Loutit, 'because they understood best the freakishness of a building cleaved open by high explosive. They knew it collapsed in one of three ways. The building might disintegrate entirely, leaving just a smoking mound of timber and masonry; the floor might cave in, forming a V and trapping people at the edges of the floor below; or there might be a curving fall of floors and roofs, one side secure while the other swung downwards.'

Rescuers were multi-skilled; they could move with extraordinary dexterity, padding across debris without sending a jet of dust into a trapped person's precious air bubble below. They could burrow their way deep into the heart of a trembling ruin, wet handkerchiefs wrapped round their mouths to protect against leaking gas. They could think with searing ingenuity, securing tunnels with table legs or using a wardrobe as a mobile prop. On Sunday morning one rescue team tunnelled into a devastated block of flats and discovered a woman called Mrs O'Leary. Her whole body save her face was pinioned under tons of rubble and the ceiling of her flat had collapsed so that it rested two feet from her head. One of the rescuers reassured her, stroked her face, and asked if she wanted anything. 'To be freed,' she laughed. He told her they were working to remove the debris as they spoke. Then she asked for a cup of tea, so the rescuer wormed his way back down the narrow tunnel and brewed her some. He returned with a cup and a sponge and for the next few minutes soaked the sponge in the tea and squeezed drops into her mouth. It was thirteen hours before Mrs O'Leary was released.

Such sensitivity wasn't uncommon among London's rescue squads. A woman stripped naked by the blast was quickly covered with an old blanket; a man who had festered in his own excrement for hours was greeted with a smile and a few friendly words. Only when he was on his way to hospital did his rescuers hose each other down and crack jokes at his expense.

WHAT saved Rose on Sunday morning was her dog. When the bomb had exploded a few hours earlier the two of them had been cuddling in the basement. She was still cradling him in her arms at first light, though she had no idea what time it was, trapped in the darkness with the smell of gas wafting around her. When the animal began to whine, Rose sang to soothe its fears. Up above a rescue worker screamed for silence. One, two, three seconds passed. Then he started tearing at the rubble like a man possessed.

At the Redriff Estate in Bermondsey 12-year-old Tom Winter had been up since dawn. On this morning he didn't feel in the mood to go looking for incendiaries. Instead he walked round the estate surveying the damage. A whole four-storey block of flats had been demolished by one HE, 'and there were two large bomb craters in Elgar Street and Gulliver Street near the Dock Alley'. Word came through that the dock policeman, 57-year-old Charlie Burton, had been crushed like a bug when a bomb landed on top of his hut in Surrey Docks. Charlie had always turned a blind eye to some of Tom's gang's pranks. Everywhere Tom looked flats had 'broken doors and windows and glass was almost everywhere on the Estate'. The sound of the glass being swept up reminded Tom of his brief evacuation to Brighton, and the sound of the sea caressing the shingle on the beach.

In the middle of Redriff Estate there was a long trestle table with watercress sandwiches laid out. From time to time one of the rescue workers trying to dig out the corpses of Granny Humphreys and the four men who died with her broke off from the work at No. 53 and walked over to the table. Each time Tom backed nervously away and watched from a distance. 'One of the men was eating a sandwich,' recalls Tom, 'and he saw me looking at him. "Do you want one, son?" he asked. I was hungry but something inside me said, "No, thank you". I think it was because he'd been digging for dead people and now I associated him with death.'

It appeared to Tom that the whole of the estate had turned out to watch them bring out the dead. But it was also a chance to swap bomb stories, to tell tales of near misses and lucky escapes, usually burnished with self-deprecating humour. Laughter was how Londoners comforted each other. People who looked for a shoulder to cry on were told to pull themselves together. 'One could panic in his heart,' reflected Eric Sevareid of the American network, CBS, 'but two together could not show it, nor a hundred in a group.'

'It was about mid-morning when the first body was brought to the surface,' says Tom. 'It was in a sack and they got it out through one of the big windows on the ground floor. The others followed one by one soon after, all in sacks.' With them came a smell, a strange smell, a new one to Tom. Later his dad told him what it was. 'The smell of death,' says Tom, 'isn't a nice one.'

The sight of death was even more repellent this Sunday. At a time when ARP wardens would normally be filing into church they were panning London's dirt looking for chunks of flesh, perhaps all that remained of someone caught by the full blast of a 1,800kg 'Satan'. Whatever was found was put in sacks or ash cans and taken to one of the improvised mortuaries where the task began of trying to identify the human flotsam.

Thomas Cockburn was one of the doctors who volunteered to help piece bodies back together again. It was different work to the prenatal clinic he ran in East London. When the contents of the first can were strewn out on the mortuary table he thought they 'looked like a load of dirty concrete just tipped out of the mixer'. Stacked against the sides of the mortuary were a dozen more ash cans. 'The first problem was to decide how many people were represented by the remains in the ash cans,' he recalled. 'The bodies had been smashed and torn beyond all recognition and splashed over the houses and gardens up to a quarter of a mile away.' The ARP wardens gave Cockburn a list of names of missing people and a brief description. Could he match them up? 'At first I thought it would be easy to distinguish male from female but even that was not possible ... the heap on the table was just a greyish brown mess, mixed with gravel, stones and dirt that had been scraped off the walls or gathered from the gutter of the houses. Legs, trunks, heads had all disintegrated. After an hour, it became apparent that the only parts of the body likely to be found in a recognisable condition were feet, so we started collecting feet.'

When Cockburn had pieced together as much of his gruesome jigsaw as possible he 'scribbled down about half a dozen lines on a report form stating what had been identified'. Soon after some more ash cans would arrive and he began another puzzle.*

MOST of London was without gas and water on Sunday morning, but by now people had adapted to such deprivations. For months they had filled their bath tubs each early evening with the regulation five inches, a private reservoir for just such an emergency. In Hackney and Peckham and Paddington the water mains had run dry. In Bermondsey, too, where Gladys Jenner and her mother had returned to their home in Clement's Road. 'We'd stayed in the guardhouse until the All Clear went,' she recalls. 'Then they told us to go home and get our heads down for a few hours.' The old tub was in the scullery and the hot water of the previous night was now as chilly as the outside temperature. With no gas to heat a saucepan of water, Gladys doused herself with cold water. 'We had some breakfast and then at around 10 a.m. we began to hear reports from last night from friends.'

Vera Reid returned to her flat in Gloucester Place filled the kettle from the bath water and put it over the fire. 'I made tea for the firemen who were still working frantically,' she said. 'They were so grateful that it made me ashamed again, for it's we who should be grateful for them.'

In Maida Vale Olive Jones's frustration was mounting. The taxi she booked at 3 a.m. had never arrived and now the 'gas was as dead

*Of the 43,000 victims of the Blitz on Britain in 1940/41, 537 were never identified.

as a doornail ... of course, it would have to be hit in a main just when after a great deal of trouble we had managed to get the year's last lovely joint of pork for a Sunday roast'. She consoled herself by basking in the 'sunny peace and beauty of the morning'. It occurred to Olive that perhaps last night never happened. Perhaps she hadn't scrambled over the wall to douse a livid incendiary in her neighbour's garden. 'The events of the previous night seemed as incredible as the fantasies of a nightmare,' she told herself. Then the phone rang. It was a friend checking she was all right. She told Olive 'that the local hospitals were crammed, their front halls running with blood and lined with bodies'. Later she heard 'that twelve people had been killed in one house and seven in another'.

Lew White phoned his wife from his sub-station in Lincoln's Inn. 'That was the first thing I always did after a raid,' he says. 'To make sure she was OK and to tell her I was fine. In some ways I was more apprehensive and frightened waiting for her to answer than I was fighting fires. My imagination began to run riot if she didn't pick the phone up straight away.' Lew's wife and 2-year-old son were safe in Willesden. 'In fact, they didn't have a clue about the extent of the damage because that part of north London hardly got hit. They could just see these big black clouds over the centre and asked what they were. I just told them we'd had a bit of excitement last night.' Wembley had enjoyed a peaceful night's sleep and now wondered why the papers were late. In the south of the city a couple out for a morning constitutional on Streatham Common chased scraps of burnt paper as they fluttered from the sky. 'The paper was falling until midday,' one of them wrote in her diary. 'The common looked like Hampstead Heath after a Bank Holiday and it was the same as far as Morden and Epsom.'

Jean Ratcliffe and her boyfriend had slipped out from the Waterloo Station lavatories at first light and gone their separate ways, she to her parent's house in Harrow and he to Epsom. 'The first thing my mother said when I arrived home was "What do you think the neighbours will say?" I could've hit her, but she had no idea what had gone on because the area hadn't been bombed.'

A civil servant from north London, Anthony Heap, had slept through the raid in a public shelter in King's Cross. His morning walk had taken him trance-like through the West End, past 'various shops burnt out in Tottenham Court Road, houses demolished in Fitzroy Square and Fitzroy Street; a good part of both Charlotte St and Old Compton St ... completely destroyed by fire; a nasty hit between Dean Street and Wardour Street ... street craters in Oxford Street & Barnard Street & Bedford Square. Fire damage on corners of Russell Sq and Tavistock Sq'. At Waterloo Bridge he looked across the river at the fires still burning in Southwark. He recorded his return journey home in his diary that night: 'In Catherine St was a crater between the Duchess and the Strand. Buildings opposite both sides of Australia house had been hit and St Clement Danes' Church burnt out. But it wasn't until I hit Fleet St that I came across the really big stuff. Huge areas on both sides of it had been burnt right out – not so much in the "street" itself as behind it. Ludgate Circus was a shambles and New Bridge St one gigantic network of hose pipes leading up from the river at Blackfriars to fight the fires still burning in Ludgate Hill, Old Bailey, and elsewhere. Ludgate Hill was in fact in ruins and nothing was left of St Bride St, Shoe Lane, Charterhouse St and the whole length of Farringdon St north of Holborn Viaduct and very little of Smithfield and the south end of Farringdon Rd. These are plain, unexaggerated facts. This whole area has been virtually laid waste...but it didn't end there ... what little had been left of Gray's Inn from the last show-down was finally disposed off. The north front of Lincoln's Inn too had taken a rap and more of High Holborn, including the stadium had been hit. The best part of Red Lion Street and Eagle Street were burnt out, Bedford Row had taken three direct hits and, as a piece de resistance, the whole of Theobald's Rd (both sides) from Bedford Row right up to Southampton Row, together with the south end of Lamb's Conduit St had been brought to the ground. The casualties must have been enormous and though the main brunt of the attack must obviously have been concentrated on this extensive area adjacent to the city, I should think the total damage sustained was as great as anything inflicted in any previous single raid. And this is the price we have to pay for so called "democracy". Is it worth it? I shall leave posterity to judge. The world is too insane today for anyone to hope to make any sense out of it all. Our water on tap again this evening.'

As more fire crews from outside London invaded the capital the firemen who had been on the front line for twelve hours were pulled back for a brief respite. They lay on their bunks, eyes red-edged and coughing violently from the smoke inhalation. Many of the sub-stations, Lew White's included, had no showers. But what did that matter on this morning? Firemen, more than most, knew London had no water.

They helped each other out of their sodden uniforms, tugging at the sleeves and pulling tunics over frozen torsos. The firemen looked ridiculous stripped naked. Black hands and faces, streaks of black running down their chests and backs. Milk white elsewhere. Most Auxiliary Firemen had only one uniform, unlike the regulars who had two or three spares, so they hung it up because they were disciplined not because they expected it to be dry when they went back on duty in a couple of hours' time.

29

BY lunchtime on Sunday London's wealthy were enjoying themselves once more. Families picnicked in Kensington Gardens and as the children played with their boats in the pond parents competed for the best bomb story. Vera Reid accepted an invitation to dine at Claridges in Bond Street but warned her host she hadn't been able to wash that morning. There were no taxis available to take her the short distance to the restaurant so she walked, past 'houses smoking and streets running with water. They were bringing out the bodies of some women from the front of one house'. Claridges was easier on the eye. 'Soft carpets, waiters with cocktails, people washed and clean ... we drank wine and saw Barbara and Martin who were married yesterday. They sat with their backs to us leaning together ... how long ago their wedding seems.' Fortunately for Vera, 'George didn't seem to mind that I was so grubby and untidy'.

A few miles east in Coutts Road in Stepney, Florrie Jenkins had found her mum and aunt. 'They'd defused the bomb by now so we went back to our house and salvaged what we could. But there wasn't much left.' Then they were sent to one of the borough's rest centres, situated in a school and staffed by volunteers. Stepney's rest centres at the start of the bombing had been shambolic, the well-intentioned staff overburdened by the number of homeless. There were too few beds, no hot food, no washing facilities and only a handful of lavatories. But when Florrie and her family arrived at their rest centre the procedure had been honed to perfection. They were given blankets, spare clothes donated from overseas, hot food and a camp bed on which to sleep. Throughout the remains of the day scores of other destitute people arrived, their lives condensed into what they could fit into a perambulator or a handcart.

It was after Sunday lunch that the Blitz trippers started to arrive in central London from the suburbs. The newspapers had a battalion of names for them: 'rubber neckers', 'Blitz Hogs', 'Blitz Tourists'. The most common was 'Gawkers', because that's what they did. They gawked at other people's misery and lived the Blitz vicariously for a few hours before returning home for tea and cakes. A reporter from the *Daily Express* gawked at the gawkers and described them on Monday's front page: 'Being Sunday and a day of rest thousands of people had nothing to do. So they came in their droves to look at the seared ruins, to block the streets, to trample on hoses, to hold up fire engines, to gawp at weary, blistered firemen, to fill the roads with their cars, to hamper the police, to stare at the grimy half-clad homeless.

'They swept through the few cafés open so that within a few hours the counters and kitchens were as bare as if locusts had been through. So when the night spotters and wardens came off duty for something to eat there was nothing ... perhaps the most foolish of all the incidents I saw was a group of people, several showing their children the sights, duck under a rope and wander down a deserted street. There was a very large sign indicating danger but they took no notice. A policeman had to run after them and order them out.'

Another newspaperman, Larry Rue of the *Chicago Tribune,* 'attempted to make an automobile trip to survey the most stricken regions ... but in many places because of craters, roped off areas or danger from sagging walls, I was forced to thread my way on foot, often for blocks at a time'. Never before had Rue seen the streets so congested the day after a raid. At times he found it hard to distinguish between 'sightseers and refugees'. There were clues. 'One beautiful girl standing dressed in only a mink coat over her shelter pajamas … automobiles forced to come to a dead stop to avoid running down pedestrians apparently still dazed and not realizing what was going on about them.'

In Mayfair's Shepherd Market shopkeepers were seeking out any undamaged goods amid the rubble of their properties like diners picking over the remains of the Sunday roast. Madeleine Henrey recalled 'two partners of a little oil shop, of which there was nothing left but a few smouldering sticks gazing down at the foundations of their once prosperous premises … their eyes were not always dry and they turned round and round and walked backwards and forwards on the cobblestones, only interrupting their miserable thoughts to exchange reminiscences'. The owner of the draper's shop arrived. The fire had been halted yards from her door. She was so grateful she laughed when someone told her that her lingerie had been pilfered by some discerning courtesans. Paillard the hairdresser tottered across the cobbles, immaculately attired with his moustache freshly trimmed. He gazed at the husk of his hairdresser's with no emotion. 'I was intending to hold a slight celebration,' he said with a bitter smile. 'I started here forty years ago this very day. I'm afraid the Nazis have celebrated for me with fireworks.'

Emily Macfarlane remembers that it took a long time to dig out the bodies from the tenement flats in Munton Road. 'They were six storey flats and the staircases went up round and round without landings on any floors. So when they were hit they just collapsed like a pack of cards. The only thing left standing was the fire stacks. Everything else was rubble.' Rescuers worked in stolid silence throughout the day bringing out bodies at grisly intervals. 'At one point they found a cat beside its dead owners,' says Emily. 'It was strange how often cats and dogs survived when humans didn't. Someone said they'd have to put the cat to sleep so we took him in. We called him "Timid" because after what he'd been through he was.'

Some of those coming into London on Sunday afternoon looked forward to the Summer Concert at the Queen's Hall. Outside the ribs of the hall they were politely informed by well-dressed gentlemen in a makeshift box office that, yes, of course the concert was still going ahead, only the venue had been switched to the Duke's Hall at the Royal Academy of Music. There had been no rehearsal and the musicians were playing borrowed instruments but that didn't matter. The audience needed music to soothe their wounded psyche. At 3 p.m. the orchestra struck up Beethoven's Overture 'Coriolan' and as the music entered the audience's soul the blackness that had hung over London began to lift.

SUNDAY afternoon in Gravesend was so idyllic some of the pilots from 74 Squadron went for a drive in the country. They knew exactly where they were going: 30 miles south to Kennington to see what they could liberate from the Heinkel shot down by Roger Boulding. 'I wanted to get hold of a 9mm Luger before the Bobbies did,' recalls Johnny Freeborn. They pulled up in the road by the crash site and whistled with appreciation. 'Either by good luck or superb skill the pilot had put his Heinkel down on its belly in a small field,' remembered Boulding. They didn't know that Richard Furthmann was dead, only that the crew 'had been taken to Chartham Hospital near Canterbury [and] we were unable to see them because of their injuries'.

Freeborn says they had little trouble talking their way past the armed guard and into the Heinkel. Freeborn rummaged around in the cockpit 'looking without success for a Luger and pinching instead the navigator's bomb sight'. He crouched down by the pilot's seat, the one Richard Furthmann had sat in only hours before, and fiddled around with the control column. One of his pals sat in the shattered glass nose of the Heinkel from where a few hours earlier Albert Hufenreuter had dropped his bombs on London and took a photo of Freeborn.

Outside the aircraft Boulding admired his handiwork, particularly 'the bullet strikes under the fuselage'. He prised out the dinghy from a panel in the top of the rear fuselage and they took it back to Gravesend. The dinghy was launched on the lake in the grounds of Cobham Hall and Freeborn retired to the 'workshop armed with a chisel determined to have a closer look at this bomb sight. I was chiselling away when I heard a voice behind me asking what I was doing. "I'm trying to get into this bomb sight," I said. He was a chap from an RAF Maintenance Unit who wanted to take it away for examination. I told him he couldn't have it and then he said "Do you know the Germans often put small explosive charges in their bomb sights for people like you?" I let him have the bloody thing'.

GENERAL George Carpenter, head of the Salvation Army, arrived in Queen Victoria Street late on Sunday afternoon, when it was finally judged safe to do so. His visit was described a fortnight later in the Army's journal, *The War Cry*: 'It was a quiet arrival, one for which words were not easily found. Salvationists had, on innumerable occasions, turned that corner, sometimes at the end of journeys round the world, to see the dear old building … now the walls were crumbled, blue sky showed through where the desks had been.' The Army had been based in 101 Queen Victoria Street for sixty years, ever since General William Booth, their founder, had moved the IHQ from the East End. Now all that remained 'were innumerable imperishable memories'.

'The group stood and watched the slowly curling smoke. Down

the streets white streams of water, forced at high pressure from the hoses, cut the brown clouds, which drifted like a fog in leaping arcs.'

A gas explosion hours after the fires had been extinguished sent water surging into the Army's basement strong room. Salvationists were now trying to rescue what records they could. 'When we managed to get into the strong room it was like going into an oven,' said one. 'There was water, knee deep, when we first entered which rose to our waists before we had finished emptying the security and deeds safe. Five or six of our chaps fainted and had to be hauled out ... to make matters worse the sewage from the broken mains penetrated the floor and you can imagine the smell!'

The fires threatening Faraday House and St Paul's Cathedral receded throughout the afternoon, beaten back by water relayed from the Thames through the new steel piping. By 6 p.m. both buildings were declared out of danger.

At Waterloo Station the flames in the vaults were close to conceding defeat after Divisional Officer Geoffrey Blackstone had hit upon an innovative strategy. He instructed the firemen to drill through the asphalt of the taxi rank to give the flames below an escape route. A dozen spumes of fire erupted from the holes and eased the pressure beneath.

Fred Cockett parked the hose-laying lorry he'd borrowed from the Westminster station hours earlier round the side of Waterloo Station. 'I phoned them later,' he remembers, 'and asked them what one of their lorries was doing in Waterloo. This voice at the end of the line said "Is it? Oh, we wondered why we were one short. We'll be round for it in the morning".'

BY 8 p.m. on Sunday night 1,000 acres of London were still without water. Only a fraction of the water mains smashed the previous night had been repaired. Over 150,000 families had no gas or electricty. The

Home Guard was out in force to prevent looting. Already in Streatham Vale, south London, thieves had taken advantage of the disorder to break into a sub post-office and remove the safe, containing £1,800 worth of money and postal orders. There were reports of men and women plundering damaged shops and drinking dry the contents of bombed out pubs. Then there were other people, like the man in St Pancras who came across an iron stewing pan lying by the roadside in which were two Golden Syrup tins. Inside the tins were bundles of £1 notes, £399 in total. He packed the tins back inside the iron stewing pan and handed it in at the town hall.

There was an emergency meeting at the Imperial War Museum to thrash out what could be done to steady the city's nerves. It was hosted by Admiral Sir Teddy Evans, Joint Regional Commissioner of London's Civil Defence, and a veteran of Captain Scott's doomed expedition to the South Pole in 1912. Then, his inane cheerfulness had exasperated Scott. Now it riled Southwark's Civil Defence Chief, Alderman Len Styles. 'Chin up, chaps!' said Evans, unsettled by a table of gloomy faces. 'In my opinion, sir,' replied Styles, 'two more nights of this and London will be at a standstill.' Evans was handed a preliminary report from the Metropolitan Water Board. At least 147 water mains had been broken. Another four more raids similar in intensity to last night's, the report cautioned, and the capital's entire water system would be exhausted.

Evans did his best to rally the troops. It was all he could do. All anyone in London could do on Sunday evening as night closed in was pray that the Germans wouldn't return. Hundreds of the 2,154 fires started nearly twenty-four hours ago were still burning as if there was no tomorrow. The Luftwaffe would have no difficulty in locating their targets. On a fireboat in the Thames George Wilkins and his crew waited. Everyone in London was waiting. For a few wretched people the waiting was too much to bear. A hairdresser in Shoreditch killed himself.

'Firemen were the only people above ground,' recalled Wilkins. The Blitz trippers had scuttled back to their leafy suburbs. Now there was just 'an awful loneliness'. The fear was palpable. Then at 9.30 p.m. Wilkins heard those 'stomach-sinking sirens wail again. We were utterly vulnerable and I remarked to my pals, "This will be our lot tonight" for we all expected a raid like the previous night'. For half an hour there was silence. Not the sound of lovers laughing. Not the sound of a motor vehicle. Not the sound of an uneven drone up above. 'Then to our great relief,' said Wilkins, 'the All Clear sounded 30 minutes later and we had a quiet night.'

30

MANY tears were shed for London in the days after 10 May. They didn't roll down the cheeks of the ardent Cockneys who had borne the brunt of the raid; nine months of bombing had calloused their emotions. The emotion came from outsiders; from Henry Wood, who broke down when he was shown the Queen's Hall; from Winston Churchill who wept as he stood amid the ruins of the Commons; from M. D. Severn, the librarian at Gray's Inn since 1895, and now in charge of nothing but the ashes of 32,000 books. Perhaps the King cried when he returned from Windsor Castle and saw Westminster Abbey and the damage to the scene of his coronation on 12 May 1937.

Whatever the emotions of the Royal Family, working-class London no longer cared. 'All this stuff about "We're suffering just like you,"' says Joe Richardson, 'what rubbish! They lived up in Windsor most of the time and when they toured the East End it just rubbed us up the wrong way. They didn't know how we lived and they didn't want to.' John Fowler remembers well the resentment felt by many Londoners towards the Royal Family. 'It upset a lot of people ... when she [the Queen] turned round and said "We're the same as you now" after an end room in Buckingham Palace had caught some bomb damage. They weren't the same as us; they had somewhere else to go. My mum and dad had nowhere.' One man, a teenager in 1940, remembers when the Queen came to the East End, 'to dance on the debris'. 'My old gran said to me "you won't see that old cow now, she'll be going up to Sandringham out the way".'

London began to stitch up its wounds on Monday, an operation that began not with a needle but with a hammer. Throughout the morning workmen hacked out the shattered glass from office windows on to the ground below, from where it was swept into glittering slag heaps on street corners.

As the surgery began, Londoners read about the raid in their morning papers. There were photographs of smoke over London and details of the damage to the House of Commons. The other stories were about the heroes of the night; the policeman who climbed up the Victoria Tower to douse an incendiary and save Parliament; the part-time firemen whose 'tempers were even [and] their nerves showed no sign of ragged edges'; the 'magnificent' RAF pilots 'who helped to destroy thirty-three Nazi bombers on Saturday night'.

Tom Finney, back at work in Preston as an apprentice plumber, 'had no idea what had happened in London until I picked up the papers on Monday morning. When I'd read them I thanked my lucky stars we left straight after the game'. Despite their coverage of the raid, however, the newspapers hadn't overlooked the cup final. It made pleasant reading for Tom. The *Daily Mail* called Finney's performance 'one of the finest features of the game' and the *Daily Telegraph* described him as a 'brilliant youngster'.

Olive Jones scanned the papers and smelled a rat. 'The papers play it so remarkably low that my suspicions have been aroused,' she told her diary. On Sunday morning a friend had phoned to say the local hospital was running with blood but the newspapers played down the human toll. Reg Harpur in Sydenham, seething at what Germany had done to Britain's history, told his diary that 'I hear casualties were very light – thirty at the most'.

Most newspapers in the English-speaking world rewarded John Hughes for his Sunday morning diligence. He told his daughter how '*The Times*, like every other newspaper and news agency in the world used my story exclusively ... the Associated Press of America, which supplies hundreds of papers in the USA cabled my story in full – over 2,000 words ... the story the *Sydney Morning Herald* cabled for its issue on the Tuesday was your daddy's word for word as he dictated it to one of the typists here at about 8.30 a.m. on Sunday morn 11 May. All the other Australian mornings used your Pappa's story, too ... If I took myself seriously I would get the World's Press News to write a lot of hooey about my scoop.'

As the papers made little mention of the damage to London other than around Westminster Palace, Olive Jones got in her car and drove into the City. 'Took an hour on Monday morning to get from Cheapside to the Strand – an hour of inching along in clouds of dust and petrol fumes, so appallingly congested were the very few thoroughfares which remained open ... but everybody, except me, accepted the never-ending wearisomeness of mile-long traffic jams with the most wonderful good humour ... most amazing sight was the pedestrians who thronged the thoroughfare. I should have hardly thought there were so many people in the world as there seemed to be office workers walking the streets carrying briefcases and type-writers and attaché cases ... Fleet St which was closed to traffic, but open to people on foot, was filled from side to side with a solid mass of walkers.'

A journalist from the *Daily Herald,* Mea Allen, toured the capital on Monday morning to report on the after-effects of the raid. 'Very weird and many of the buildings still blazing, flames licking round windows frames,' she wrote to a friend. 'I left Golders Green in the morning in brilliant sunshine; stepped out at Leicester Square station into a grey overcast day, leaden clouds above and visibility bad – like a November fog threatening to come down and envelope the city. The sun did not break through the smoke pall till 12.30 ... all morning we had a black snowstorm – burnt out paper, etc, drifting down. It was incredible – every street littered with it. I have never seen such widespread damage.'

South of the river it was pandemonium. One commuter, Doug Bathurst, embarked on his daily journey into work swearing blind that Hitler wasn't going to stop him reaching his City office. 'I left home at 7.30 and got a train from Streatham to Herne Hill where we changed to relief buses and set out for Holborn via Denmark Hill, Camberwell Green and Camberwell Gate, through various side streets, across Newington Butts to Kennington Lane where the traffic jam was hopeless owing to all the traffic being diverted to Lambeth Bridge ... during the bus ride I could not even count the craters, wrecked houses, burning buildings and roped off streets.' Seeing that this road was going nowhere, Bathurst hopped off and carried on on foot. 'Walking up Blackfriars Road I found the Eye Hospital damaged, the block next to it [Surrey Music Hall] a smoking ruin and several other buildings still burning. The [Blackfriars] bridge was closed to all except pedestrians and cyclists. The crowd walking through New Bridge Street dodged the spurting water from hose pipes and stared at the firemen pouring water on the ruins of Ludgate and its side streets. The same was happening all along Farringdon Street and past Holborn Viaduct I found a whole block gutted and Charterhouse Street impassable. Several large bombs in Smithfield Market and some of the fires were still burning this morning.' Doug made it into work at 10 a.m. and found the City without water, gas or electricity.

Jean Lovell's train from Chadwell Heath in Essex pulled in to Liverpool Street several hours late. Her walk west towards the offices of the carpet manufacturers in Gresham Street where she worked took her through a distorted City. 'The devastation was horrendous,' she recalls. 'There were firemen lying by the roads absolutely exhausted with their hoses everywhere. My dear old firm was gutted so I returned home. That afternoon I received a message telling me manufacture had been transferred to our Uxbridge factory and that the next morning a van would pick up me and the others from Liverpool Street and take us to the new office.'

The transport network had been shredded right across London. For twenty-four hours after the raid Marylebone had been the only mainline station running a service. A Camberwell couple who had stolen a weekend together in Northampton were turned off their train to Euston at Kentish Town on Sunday evening. Beyond this point the track had unravelled and with no buses heading south they had to walk the seven miles home. Euston was reopened on Monday, as were King's Cross, Liverpool Street and Paddington. But Victoria and Waterloo remained closed for a week.

The destruction to the London Underground was immense. Nearly 30 miles of track had been destroyed and there was a UXB outside Moorgate Station. That was defused on the Tuesday but other damage took longer to repair. Aldgate Station didn't reopen until 21 May; the service between St James's Park and South Kensington resumed on the same day; the stretch of track between Rotherhithe and Surrey Docks was closed until 8 June and the Circle Line from Baker Street to King's Cross only reopened on 21 July.

The situation on the roads was appalling. Just under a thousand, 959, were closed on Sunday and the police had soon exhausted their supply of yellow tin 'Diversion' signs. Craters left 579 roads resembling a First World War battlefield, with 315 impassable because of debris or buildings with drunken angles tottering over them, and sixty-five rendered unusable by UXBs. On Monday all but two of the bridges – Tower and Lambeth – were still blocked. The bomb that had exploded through the north side of Waterloo Bridge had caused serious damage to the Kingsway Tramway Subway and there were no trams or trolley buses in to Westminster from the southern side of the city.

No buses were operating in the City on Monday and it wasn't until Blackfriars Bridge reopened on Tuesday that a link was re-established across the river. Two bus garages had been hit, in East London and in Croydon, with the loss of four employees and 113 buses.

The greatest disruption to the transport network were unexploded bombs (UXBs). A post-raid intelligence report stated that a total of 2,932 high explosives, 72 oil bombs and 77 parachute mines had been dropped during the raid; of these 162 had not exploded. As well as the UXBs at Victoria and Moorgate stations, there were ones outside Clapham South, Chancery Lane and Goodge Street stations. The 10ft parachute mine suspended above the stage of the Palladium was defused on Sunday by a naval bomb defuser who, for his courage, was rewarded with free tickets to the theatre for life.* 'We didn't have a show on Sunday,' says Vera Lynn, 'so when we came in to the theatre on Monday the bomb had been removed but the parachute was still there. We cut pieces off the green silk not with the intention of making dresses, just as souvenirs.'

A large section of London's industrial output had been battered into dormancy. A firm of scientific instrument makers and refrigerating engineers in King's Cross were put out of action for a fortnight; the Waste Paper Company in Waterloo had been destroyed and spewed burned fragments over the capital for three days; production was also disrupted for a significant length of time in a tin-foil factory and an engineering plant in Southwark, an Iron Foundry in East Ham, a Printing Works in Camberwell, a sugar refinery and food production works in Poplar, a firm in Stepney that made Bailey Bridges. The tailoring trade in the East End was brought to a standstill by the loss of

* As parachute mines were originally sea mines they were the sole responsibility of naval bomb disposal squads.

electrical power; in Bermondsey output had been suspended in a dozen factories including Peek Freans, who were unable for several days to resume production on its ration packs. In the Elephant & Castle, Dean's Rag Books, which had diversified into Mae Wests for airmen, was burned down, and in Pimlico the Palmolive Soap factory in Ranelagh Road was no more, though the fragrance left by its cremation masked the stench of death that hovered in the air as rescue workers a few hundred yards away pulled out twenty-four bodies from the Turner Buildings in Millbank.

But the one factory that the Germans wanted to hit above all others had barely been touched. Several Luftwaffe navigators had picked out the Royal Arsenal in Woolwich, lying on the south bank of the Thames just before the U-bend, but none had found its mark. The first rash of incendiaries freckled some open ground behind the Arsenal at four minutes past midnight. Inside, the mostly female nightshift workers had headed for the shelters. Each one had a number painted on her overalls. If anyone was seen running for cover, their number was noted, and they were sacked on the spot. One must never run in the Royal Arsenal – even when more incendiaries landed at 12.30 a.m. or when the first high explosives destroyed the guard hut at 2.20 a.m. Twice more during the night the country's biggest arms manufacturer was targeted, but the bombs fell on a roadway and on an empty air raid shelter.

Sixty other priority factories unmolested by the raid were impotent on Monday because they had no gas. Twenty more suffered low pressure. The destruction of the South Metropolitan Gas Company's station holder in Kennington had deprived virtually the whole of south London of gas. In Poplar the damage sustained by the Commercial Gas Company had stripped much of the city of its supply. The biggest problem for the army of gas engineers was the invasion of broken gas mains by water that poured out from splintered mains. In hundreds of homes men and women who had turned on their gas stoves stood wide-eyed and dripping wet as water fountained into the air. Pumping out the water took a long time. On 17 May, 120,000 homes across London had yet to have their gas supply reconnected.

Elsewhere on Monday morning, there were 237 main telephone cables out of order, including 53 trunk lines and 131 junctions. Worst affected was Ludgate Circus, which was without 69 of its cables. In Westminster the electricity supply was in chaos following the crushing of the St Martin's Lane sub-station, and another explosion in a cable chamber off Oxford Street. Eighty per cent of the outgoing cables were damaged.

The Metropolitan Water Board soon revised its initial estimate of 147 water mains broken. The new figure was 605, of which 98 were 12ins in diameter or over. A fleet of over 1,000 water tank wagons began to trundle through the streets as water board engineers strove to repel the hidden enemy, typhoid.

A small outbreak of the disease in 1937 had killed forty-three people in Croydon and left a residue of fear in the population. A pre-war report by the Metropolitan Water Board warned that should a filtration works become infected with sewage during a raid there would be approximately 16,000 cases of typhoid, 1,600 deaths and '800 permanent carriers of the disease as a reservoir of infection for the creation of further outbreaks'.

The water board had implemented several emergency measures in anticipation of the Blitz, though none was as vital as their 'pre-chlorination' programme. The chlorine in London's water supply was increased until its taste verged on the intolerable. A few people complained but most bore it stoically when they were told of the reasons. 'A bomb would often burst a water main and a sewage pipe,' recalled Arthur Durling, one of the board's engineers, 'and a certain amount of sewage would get into the mains. But we had at our depots tins of chlorine powder and a chart that showed us the amount of powder we had to put into a fractured mains when we repaired them.'

Durling and his colleagues mixed up the powder in a bucket and poured it into the hydrant close to the charging valve. 'Once you'd poured it into the main you shut it down and let the water push the chlorine along ... after about 20 minutes or so you flushed the main of the chlorine.'

For the bigger mains, 12ins and over, the fractures were repaired and then portable gas chlorinators fed the chlorine into the mains. 'Once that was thoroughly washed through and sterilised,' remembered Durling, 'the main was not put back into service until samples had been taken and tested at laboratories [for bacteria].' Most of London's water supply was running normally by Thursday 15 May. 'There were no cases of typhoid during the Blitz in London,' said Durling. 'I think that was quite a record ... [and] we felt very proud of that because a serious disease would have caused a lot more damage to the people than the whole Blitz.'

THE leviathan craters that defaced so many of London's roads on Monday teemed with life. To Olive Jones and the city workers battling to get to work it might have looked like a rabble but there was a strict hierarchy in place in each crater. It was the job of the representative of the City Engineer to decide who took precedence: was it the gas engineer, the hydraulic power expert, the sewage repair team, the water board, the engineer wanting to mend the telephone or the squad from the electricity company? All were often called out to operate on a crater's wounds.

They couldn't start work until the craters had been cleared of debris, and on Monday morning no one much fancied climbing down into the craters and removing the rocks and the mud and the clay. A Civil Defence report on the aftermath of the raid drew attention to the 'youths aged 18–20 declining to accept employment ... on debris clearance with wages at 1/6½d, which was higher than that offered for vital production work'. When the crater's innards were eventually cleaned it was almost always the sewage team who jumped in first. Sewage culverts were deepest and needed new bricks laid. Then came the electricity cables and the hydraulic power mains. The gas mains were sunk lower than the hydraulics but because they were made of heavy iron they waited until everything else was secure. The penultimate operation was performed by the water board, mending the main and sterilizing it. Then the road was tarmaced and traffic could begin to flow again. The whole process required patience. In Fleet Street the gas didn't return until Monday evening and it wasn't possible to make outgoing phone calls until Wednesday. Traffic wasn't able to pass along until Friday 16 May and Ludgate Hill was barred to traffic for a further two weeks. Worse affected was the Temple, just a few hundred yards south. The water was reconnected by lunchtime on Sunday; the electric was back on eight days later. But they remained without gas for five weeks.

ON Monday afternoon Arthur Greenwood, Minister without Portfolio, took a select group of journalists on a tour of Parliament. His job was to plan the rebuilding of London after the war. 'This too,' he indicated with a sweep of his hand, 'will now come within my province and I must think about its reconstruction.' They entered the Commons Chamber from the door behind the Speaker's Chair. The ashes from the chair blackened their shoes as their footsteps scrunched across the wreckage. The Press Association lobby correspondent described in his report how he 'looked at the unrecognisable mountain of rubble and ruin, which, four days ago, was the thronged Debating Chamber, pillars, leather-backed benches, fine vaulted roof, Diplomats' Gallery, Ladies' Gallery, the Distinguished Strangers' Gallery, soaring windows and oaken walls were all piled in a mass of masonry, 50ft high. It still smouldered in the sunlight. We tripped over the remains of the great candelabra of clustered lights that had hung from the ceiling.'

On the other side of the Chamber there was just a formless gap in place of the stately entrance through which hundreds of MPs had passed on their first day, bowing nervously to the speaker as they advanced to take the oath.* 'The great black doors,' wrote the PA's lobby correspondent, 'which were slammed by ancient tradition in the face of Black Rod when he came to the Commons, have vanished. They are now somewhere among the boulders and the bricks where the Speaker's Chair, the famous despatch boxes, the cross benches and everything that was Parliament are lying.'

The parliamentary correspondent from the *Daily Telegraph* scaled the wreckage of the Chamber and dropped down into the 'No' Division lobby. 'Four chairs stood forlornly around a blackened table. Their leather had melted but on the backs of them a gilt portcullis still showed. Among it all stood a heavy silver inkstand and paper rack stuffed with half-burned notepaper stamped with the House of Commons mark.'

On the bright side, Greenwood told the shaken journalists, the mace had been saved, as had the Prime Minister's private room and the library. Percy Carter, parliamentary correspondent for the *Daily Mail*, struck a similarly upbeat note in his copy. 'To those of us who have worked at Westminster for so long, it is sad to think upon the fairness which has been wantonly destroyed by peevish people. But let us not waste time on sentiment. On with the war. All wounds will be healed in the new world we build.'

One of the correspondents commented on the incongruity of the ruined Chamber with the fact that every clock in the Commons was

* Two brass bound boxes were kept in front of the Speaker's Chair, one containing the oath and the other a Bible. During clearance work the Bible was found to be undamaged.

accurate even though many were scarred and begrimed. Perhaps they had taken their lead from Big Ben, whose 14ft minute hand ticked on imperiously, to within one and a half seconds seconds of GMT, dismissive of the hole in the gilded fringe below his face. The previous day BBC engineers had spent a frantic few hours at a Westminster Palace without electricity installing dry batteries to provide a current so that the chimes of Big Ben would be heard at 9 p.m. that evening, the time when its nine strokes heralded the start of a 'Silent Minute's reflection on the continuing struggle for Freedom, Justice and Unity'.

At Westminster Abbey the Dean, Dr Paul de Labillière, was being consoled by Henry Morton, a Fleet Street veteran, and the journalist who had broken to the world the news of the discovery of Tutankhamun's tomb in 1923. 'He was as gentle and as courteous as ever,' wrote Morton. '"I have nothing except the clothes I stand up in",' he said, as he asked me to go and look at the ruins of the Deanery. Passing through the Jerusalem Chamber, we paused at the edge of a pit that was open to the sky, where oak beams were still smouldering, and watched the wind agitating a hideous brittle mass of blackness which had been the Dean's library. "I think I am the only Dean in history," he said ruefully, as he gazed towards the ruin of his beloved books, "who has no Bible or Prayer Book."'

Morton later ran into Jocelyn Perkins who, having had time to absorb the sacrilege to his Abbey, was in no mood for forgiveness. 'I can imagine the storm of anger that will sweep over the Dominions and America when the news of this latest exhibit of Aryan culture reaches them,' he fumed. 'The story ought to be told and I hope it will be told.'

Throughout the day, MPs arrived at Westminster to retrieve what they could; they left with books and papers from their lockers stuffed into their briefcases. Others came to pay their last respects to the

Commons, including the Liberal MP Leslie Hore-Belisha and Henry Channon. When Channon arrived the crowd of sightseers milling around outside was so dense he couldn't force a way through into the Palace. Then he bumped into Hore-Belisha and the two went for a walk along by the river. Channon wrote in his diary that what Hore-Belisha had seen had convinced him that Britain needed a new leader with new ideas. 'The country would soon wake up and realise that speeches were not victories he said, and that we were drugged with Winston's oratory. He is gloomy about the future; and sees little hope if we continue as we are now doing.'

Late in the afternoon Churchill was driven into Palace Yard. He stepped out of the car dressed in a black overcoat and slouch hat, a wreath of cigar smoke around his head. With him were Lord Beaverbrook, Minister of State, Lord Reith, Minister of Works and Building, and his wife and one of his daughters. The tears Churchill had shed in the Commons were gone when he left a while later, cheered on his way by a large crowd that had gathered.

A few hours later Churchill drafted a statement that was issued at 11.20 p.m. on Monday evening. It was in response to a German communiqué promulgated two hours earlier about a missing Nazi. 'Rudolf Hess, Deputy Fuehrer of Germany, and Party Leader of the Nationalist Socialist Party, has landed in Scotland under the following circumstances. On the night of Saturday, 10 May, an ME110 was reported by our patrols to have crossed the coast of Scotland and be flying in the direction of Glasgow ... shortly afterwards a German officer who had baled out was found with his parachute in the neighbourhood suffering from a broken ankle. He was taken to a hospital in Glasgow, where he at first gave his name as Horn, but later on he declared he was Rudolf Hess.'

In a stroke Churchill had lifted the morale of the country. Newspapers cleared the front pages to cover the story for their Tuesday editions.* 'Hess Flees to Britain: Hitler's deputy bales out of plane over Glasgow' was the *Daily Mail's* headline. The *Daily Telegraph* ran with 'Hitler's Deputy flees from Germany: Nazis seek to prove he's demented'. The *Daily Mirror* chose 'Hess, Hitler's Deputy, Lands in Britain'.

The aftermath of Saturday's raid was relegated to the inside pages: criticism of firewatchers for not doing their job, the damage in Parliament; Churchill's determination not to bend, and the hurt caused to the Luftwaffe. '124 in 10 nights' trumpeted the *Daily Express* headline, explaining that in the first ten days of May the Germans had lost 124 aircraft to RAF nightfighters.

There was still no clue offered as to the human cost of the raid, though the *Mail* said assuredly that 'fatalities were small considering the number of high explosives raining down'.** The censor forbade such speculation. Londoners, however, knew there had never been an air raid like it. Olive Jones's postman told her 'My place is smashed to smithereens – not one stone left standing on another, but Good Heavens, it's not worrying me; you feel it's wonderful anything's left at all, including yourself.'

Londoners knew there had never been an air raid like it because the postman was just one of thousands of people swamping the rest centres. They knew it because houses, entire streets, had been destroyed. They knew it because after four days the Civil Defence began using chlorine to dispel the stench from decomposing bodies still trapped in the wreckage. They knew it because each of them had heard of someone killed or wounded. What they didn't know were the precise figures. They didn't know that 11,000 houses had been

* John Colville, Churchill's Private Secretary wrote in his diary 'The Hess story has, of course, made everyone gape ... in the Travellers [his club] at lunch-time there was no other topic.'

** The Government censor controlled with a strict hand all casualty statistics. Obituary notices in the papers were forbidden from giving the actual date of an air raid victim; only the month of death was permitted.

damaged beyond repair; they didn't know that 12,374 people were homeless; they didn't know that 1,800 had been seriously wounded; and they didn't know that 1,436 people had been killed.

Never in the city's 1,898-year history had so many Londoners perished in one night. And the final death toll didn't include those who died in the days and weeks that followed; not the demolition worker who was crushed to death by falling masonry in the New Kent Road; nor the 25-year-old canteen worker in the Peek Freans factory who gassed herself and was described by the coroner as having become 'imbalanced' by the bombing; nor the manager of a building company in Thornton Heath who drank three ounces of spirits of salt on 17 May because, so his brother said at the inquest, 'he was worried by the air raids and could not sleep at night'. One month after the raid the Reverend William Pennington-Bickford, rector of St Clement Danes, died aged 67. His wife said his heart had been broken the night he watched his church burn down.

A year later, in June 1942, London had recovered. There had been no heavy raids on the capital for over a year. The night of 10/11 May had been it, the savage climax to the Blitz. Londoners had been unable to believe it at first. Each night they had waited for the uneven drone of the bombers, but they never came. Only on 21 June 1941, when the invasion of Russia began, did they understand that their resilience had won through. Hitler had realized he could not shatter the spirit of the British.*

By June 1942 people had stopped sleeping on the Underground platforms. Anderson shelters were rarely used and the only people to slink into street shelters were amorous young couples in frantic need of privacy. And it had been weeks since the banshee wail of the siren.

The bones of many buildings continued to be carted away as London put itself back together. Around the Elephant & Castle the

* The very last raid of the Blitz was a 111-bomber attack against Birmingham on 16 May.

clear-up operation seemed never ending, so demolition workers worked late, picking up overtime. On Saturday 6 June a row of ruined houses on Gurney Street, off the New Kent Road, toppled on 10 May by a stick of HEs, was cleared. The men packed up in the early evening and went off for a pint. The local children moved in, playing cowboys and Indians in the street. Just after 9 p.m. Emily Macfarlane and her mum left their flat in the Peabody Estate to go and have a drink in the Elephant & Castle pub. It was still daylight. As they walked down Gurney Street Emily's mum changed her mind. She didn't fancy the Elephant tonight. Why not the Doctor's Pub in Walworth Road? They turned round and set off back up Gurney Street. Two minutes later a terrible explosion rocked the ground under their feet. Emily recognized what it was immediately, by the noise and by the way the pressure of the blast tried to squeeze her lungs out of her body. There was silence for a few seconds, then bricks began to rain down, thudding off the dead and dying and dropping into a crater 35ft wide and 7ft deep.

Of the eighteen victims, six were children under 9 years old. Three were from the same family. An investigation concluded that the UXB had fallen in Gurney Street early on 11 May 1941 and been overlaid by debris from the stick of HE bombs. The workmen clearing rubble in Gurney Street on the Saturday had reactivated its clockwork fuse during their digging. The number of dead from London's longest night had risen to 1,454.

Epilogue

'ON the night of 10 May 1941, with one of the last bombs of the last serious raid, our House of Commons was destroyed by the violence of the enemy, and we have now to consider whether we should build it up again and how, and when.' Winston Churchill stood before Parliament in its temporary accommodation in the House of Lords' Chamber. It was October 1943 and the Prime Minister moved that a select committee should be appointed to consider plans for the rebuilding of the Commons Chamber. He continued: 'We shape our buildings and afterwards our buildings shape us. Having dwelt and served for more than forty years in the late Chamber, and having derived very great pleasure and advantage therefrom, I, naturally, would like to see it restored in all essentials to its old form, convenience and dignity. I believe that will be the opinion of the great majority of its Members. It is certainly the opinion of His Majesty's Government and we propose to support this resolution to the best of our ability.' Those who demanded the Chambers be enlarged to accommodate all MPs were, in the eyes of Churchill, gibbering. 'A Chamber formed on the lines of the House of Commons should not be big enough to contain all its members at once without overcrowding, and there should be no question of every member having a separate seat reserved for him. If the House is big enough for all its members, nine-tenths of its debates will be conducted in the depressing atmosphere of an almost empty or half-empty Chamber ... there should be on great occasions a sense of crowd and urgency.'

Thus the architects clamouring to rebuild the Commons had their unofficial brief. For the next eighteen months Parliament considered various submissions until, on 25 January 1945, Churchill and the House agreed to a plan submitted by architect Sir Giles Gilbert Scott and engineer Dr Oscar Faber. Gilbert told Parliament he estimated the work would occupy six years at a cost of £784,000 (in 1939 prices) and that the Chamber would be rebuilt to Charles Barry's original dimensions. In the interim, Parliament would continue to sit in the Chamber of the House of Lords with the Lords in the semi-adjacent King's Robing Room.

The new Chamber was opened on 26 October 1950. The final cost of £2 million was more than double the original estimate and the architect had defied Churchill in some of his reconstruction. The floor of the chamber was the same size as its predecessor, 68ft by 45½ft, and the members benches were upholstered in the familiar green hide with sound amplifiers in the backs. But above the gallery level Sir Giles had expanded the Chamber from 46½ft by 84ft to 48ft by 103ft, providing seating space for 171 additional reporters and spectators. The space above and below the Chamber was also put to good use with sixteen Ministers' rooms and two Ministers' conference rooms on the ground floor below and numerous offices above.

The new Chamber was furnished by the Commonwealth: the new Speaker's Chair (a temporary one had been in use since 12 May 1941) was a gift from Australia, made of black bean from Queensland; New Zealand donated the two despatch boxes, Canada the Table, the three clerks' chairs came courtesy of South Africa and the entrance doors were given by India and Pakistan.

The entrance from the Members' Lobby was renamed the

'Churchill Arch' (it was created from the rubble of the former chamber) and the former Prime Minister marked the occasion with a speech. Earlier in the month he had celebrated the fiftieth anniversary of his election to Parliament.

In 1969 a bronze statue of Churchill was unveiled in the House of Commons, just to the left of the arch that bears his name. After his death four years earlier, Churchill had lain in state for three days in Westminster Hall, the first non-royal since William Gladstone in 1898 to be so honoured. Thanks to the alacrity of Walter Elliot and the bravery of London's firemen, the Hall's 660-ton hammer-beam roof formed a respectful Guard of Honour over the coffin. 3,000 cu ft of wood was needed to repair the damaged roof. Australia offered to supply it but Parliament preferred English oak and the timber came from the Wadhurst estate of the Rt Hon. Sir George Courthope.

The damage to Big Ben was superficial. The glass on the south dial was replaced and it was found that whatever had struck the clock – an anti-aircraft shell was the most likely culprit – had not affected its accuracy. With wonderful irony, however, one of the workmen repairing the damage to the belfry dropped his hammer into Big Ben's mechanism a few weeks later and stopped the clock.

ON Sunday, 18 May 1941, Archdeacon F. L. Donaldson had addressed a large congregation in Westminster Abbey. Alan Don was mesmerized by 'the light streaming downwards from the open Lantern ... [it] has a beautiful effect'. In his sermon Donaldson told the congregation that 'this church was erected not only as a place for prayers to be said in but as a symbol of the beauty of God – and that symbol still exists. The damage is material damage only'.

The Coronation Chair returned to the Abbey from Gloucester Cathedral in September 1945 and by March the following year the rest of the treasures were back. The nave altar was used while the

Lantern roof was repaired and the restoration met with general approval. The Victorian stained glass that had been damaged was replaced by plain glass, and chandeliers of Waterford crystal were introduced. The Deanery and the Deanery courtyard were rebuilt by Seely and Paget, who undertook much of the repair work across London between 1946 and 1956. The houses in Little Cloisters (where Dr Perkins still lived at No. 5) were rebuilt between 1950 and 1958, though No. 3 is not on the same site and No. 6 is a new building. In the official history of Westminster Abbey, Edward Carpenter wrote that: 'All things considered the Abbey and its precincts were in better shape after the war than before it.'

Westminster School was rebuilt with the school architect Mr A. L. M. Russell overseeing the work. 'It is proposed to use brick and not to make any effort to reproduce the style of our old buildings,' he said prior to embarking on his task. 'Our aim is to erect completely new buildings which, although modern, will fit in very reasonably with what remains of the old school.' King George VI opened the new school buildings in 1950.

Canon Alan Don's church, St Margaret's, accommodated the congregation from St John's in Smith Square after its destruction on 10 May. In 1943 the Rt. Hon. Captain S. A. Fitzroy, Speaker of the House of Commons since 1928, was buried beneath the floor of St Margaret's chancel and the grave overlaid with a stone from the floor of the wrecked Commons Chamber. The church escaped damage during the 'Little Blitz' of 1944 and the following year it hosted Parliament for a Service of Thanksgiving to mark the end of the war. The lesions of the Blitz can still be seen, however, sixty years later, at the eastern end of Pew 38 where incendiaries once landed.

St John's was restored after the war and since 1968 has been one of London's major concert venues, famed for its exquisite acoustics. Across the river in Lambeth Palace, for a long while only the basement was fit for habitation. Over 3,000 books had been destroyed in the library and the top two floors of the Lollards' Tower were burned out. An inspection of the Tower (originally built in 1435 at a cost of £291 19s, by skilled workmen earning six shillings a day) found that the medieval boarded lining presented difficulty for restoration as the rings to which the prisoners had been chained were firmly fixed into the wall behind the boarding.

The reparation of the Palace was undertaken by Seely and Paget, the same men who had repaired Westminster Abbey. The Lollards' Tower needed comparatively little work and the refurbishment of the red-brick Gothic Hall was received with general approbation. The Chapel, however, rebuilt in 1955, was described by one critic as 'bland and chilly'.

ST Nicholas Cole Abbey in Queen Victoria Street had a brief foray into show business after the war. The 1951 film *The Lavender Hill Mob*, starring Alec Guinness and Stanley Holloway, used the crippled church as a backdrop to a gold bullion heist. Only a few years later did restoration work begin, with diligent adherence to Christopher Wren's original design. The upper part of the tower was rebuilt and the spire heightened. From 1982 to 2003, the Abbey was leased to the Free Church of Scotland. Since then it has stood empty while the Church of England ponders its fate.

The mammoth task of rebuilding St Mary-le-Bow in Cheapside began in 1956, under the direction of architect, Laurence King. He based the restoration on Wren's structural design nearly 300 years earlier. The steeple was dismantled so the foundations could be strengthened, and when the bells rang out again in 1961 only a few Cockneys detected they had been recast a semitone lower.

All that remained of St Mary's Newington after the raid was the tower. The bells lay in the scorched guts of the church, but not for long. 'Someone has stolen all the bells,' wrote Joan Veazey in her diary in June 1941. 'They were lying there for all to see, yet no one saw them disappear!' Over the next few years services were held in the adapted church hall. The new St Mary's wasn't consecrated until 1958, the foundation stone having been laid by Princess Margaret the previous year. A new rectory was built behind the church where the current vicar resides.

St Clement Danes was rebuilt with money from an RAF appeal fund and a bequest from the wife of William Pennington-Bickford. She died three months after her husband and on the ruins of the church was laid a brass plaque:

'Near this spot rest the remains of William Pennington-Bickford, beloved rector of St. Clement Danes for 31 years, died June 12, 1941, from shock and grief caused by the destruction of this church by enemy action, also of Louie, his wife, who died September 5, 1941. They were lovely and pleasant in their lives and in their death they were not divided.'

The plaque was lost during the rebuilding of St Clement Danes, which was completed in 1958. The Queen and the Duke of Edinburgh attended its reconsecration in 1958 as the central church of the RAF. The badges of over 700 squadrons adorn the floor tiles and a roll of honour listing all the force's fallen during two world wars is kept in leather-bound books either side of the altar.

LIKE Pennington-Bickford, Henry Wood was rendered distraught by the vandalism of 10 May 1941. Yet in spite of the violation he felt at the demise of the Queen's Hall he was unable to loosen the bonds. Throughout that summer Wood returned to stand atop the slag heap of ruins and listen in his mind to Richard Strauss and Marcel Dupré, Dame Nellie Melba and Arturo Toscanini. Wood described one of these visits in a radio interview on the BBC three months later:

> I went to see it again some days ago. Round to the old familiar door. There ought to have been a platform at the end of the passage, and rank above rank of seats, and an organ towering up. There was just a mound of ashes strewn with glass and heaped with rusty iron. The only roof was the sky. There wasn't a sound. I turned to go; but as I turned my foot caught something. I knelt down and groped among the ashes. There it was, hard and circular like a plate. A cymbal, that's what it was. All in the way of music that enemy bombs had left of Queen's Hall. A cymbal. But that's significant. A cymbal is among the oldest musical instruments known to man. You'll find it in the Bible. And even now, nothing can take its place in the orchestra. It's trumpets for a victory, but the cymbal for a triumph.

The Proms were transferred to the Royal Albert Hall in 1941, a temporary measure that grew into permanence, but Henry Wood never lived to see Britain triumph over Germany. He died, aged 75, in August 1944, a month after he had conducted the opening night of his fiftieth Promenade season. The Queen's Hall was never rebuilt and for all but the oldest Britons it is the Royal Albert Hall that is synonymous with the Proms.

The Queen's Hall was irreplaceable to many, but other concert halls emerged in post-war London; and in the most unlikely places. By the Thames at Waterloo, where Fred Cockett had pumped water to help save his baker's, the South Bank started to take shape in the 1950s. For centuries there had been a congestion of wharves and warehouses along this part of the embankment, and plans had been afoot since the mid 1930s to transform the area, described by the LCC as 'depressing' and 'lacking in dignity'. Work began in 1948 and the first building to be unveiled was the Royal Festival Hall in 1951, followed, subsequently, by the National Film Theatre and the Queen Elizabeth Concert Hall. Now there are galleries, restaurants, the National Theatre and, the latest addition in 2000, the Millennium Wheel.

In 1950 it was decided to abolish London's trams. The tunnel on the north side of Waterloo Bridge, beneath Kingsway and Aldwych, was closed on 5 April 1952. 'Last Tram Week' began three months' later and on 5 July the last one ran from Woolwich to New Cross. A stretch of the Kingsway Tunnel was opened to motor traffic in the 1970s with a rampway just south of Holborn, but the old exit a few hundred yards north lies dormant behind heavy padlocked gates. Trolleybuses survived until 1962 when they were replaced by buses.

East of the Aldwych in Fleet Street the newspapers have all decamped to less intoxicating climes. The exodus began in 1985 when *The Sun* (successor to the *Daily Herald*) and the *News of the World* moved to Wapping in East London. The hot metal of the printing presses had ceded power to new technology and it was cheaper to produce newspapers outside central London. Other newspapers followed and law firms and businesses moved in to the abandoned offices. The last to go was Reuters (where John Hughes had written one of the scoops of the war), which sold its premises at 85 Fleet St for £32 million in 2003 and moved to Canary Wharf.

The four Inns of Court on either side of Fleet Street suffered varying degrees of damage on 10 May. Lincoln's Inn rode its luck and survived almost intact; some offices were scorched and there was a fire in one of the Stone Buildings, but the library's 70,000 books were protected by Lew White and the firemen of sub-station A1Y.

Half a mile north of Lincoln's Inn, the aching desolation of Gray's Inn was overwhelming. In 1945 young trees were sprouting in the severed ruins and some questioned whether it should be rebuilt. King George VI initiated the Inn's revival with a donation to the library of a complete set of Statutes at Large. On New Year's Eve 1945 a temporary prefabricated library was opened at Gray's Inn by Winston Churchill, who said it was 'the architecture of the aftermath'. The prefab remained until the restoration of the original library in 1958. Francis

Bacon, shaken and a little scarred, was put back on his plinth. The chapel was completed in 1960 and stands still on the same site as the chapel provided for John le Grey in 1315.

The years of 1942 and 1943 were but a brief respite to the other two Inns of Court, the Inner and the Middle Temple. More incendiaries were dropped in March 1944 and three months later a Doodlebug exploded near the Temple Underground station, blowing out every window of the Inns of Court. 'In the five centuries of its history the Inner Temple has suffered many times from fire and destruction,' wrote the sub-treasurer Sir Frank Mackinnon in August 1944, 'and what has been lost has been restored and rebuilt. The task faces us on a greater scale, but we shall achieve it'.

It was another fourteen years before the two Inns of Court, and the Temple Church, were restored to their former glory, during which time the original architect was sacked for being too slow. The cost of rebuilding the Inner Temple alone cost £1.5 million pounds (in 1950s prices). The War Damage Commission provided £1.4m and the other £100,000 came from the Inn and generous donations from the American and Canadian Bar Associations. On 7 November 1958 Queen Elizabeth II attended the rededication of the Round in the Temple Church.

The only single building damaged on 10 May to need longer in its rebirth than the Inns of Court was the International Headquarters of the Salvation Army (IHQ). After its obliteration on 10 May (the two hundredth building belonging to the Salvation Army to have been destroyed in the Blitz) the Army was put up temporarily by the Canadian War Services Hostel in Holborn. A few weeks later IHQ moved into new lodgings south of the river into the Salvation Army's Training College in Denmark Hill. There it stayed until 13 November 1963 when Queen Elizabeth, the Queen Mother, reopened the new IHQ at 101 Queen Victoria Street. In 2001 the Salvation Army made the same journey across the river to Denmark Hill, as the IHQ premises were redeveloped once more. This time, thankfully, the stay lasted three years not twenty-two and the Salvation Army is now back in Queen Victoria Street, where its founder, William Booth, had first led them in 1881.

Waterloo Station has changed little in sixty-four years, though during rebuilding work in York Road in the 1950s an unexploded 1,000kg bomb was unearthed and safely defused. The public lavatories that men and women were ushered into when the Luftwaffe arrived over London on 10 May are as deep and as cavernous as ever, and the famous clock ticks on looking down on the commuters scurrying to and from work. Perhaps, a cynic might scoff, the only difference is in the trains. They were more reliable during the Blitz.

Fred Cockett's baker's in The Cut has gone, but there is still an eel and pie shop, a throwback to another age. Before the war the street was described as 'filthy and forbidden'; now it's a haunt of trend-setters. The Old Vic Theatre was joined in 1970 by the Young Vic, an experimental theatre created by Lawrence Olivier for angry young talent. There are also radical bookshops and vegetarian restaurants. The Cut's changed.

The Elephant & Castle is unrecognizable now to how it looked before 10 May 1941. Spurgeon's Tabernacle is still there, so too the Bakerloo Underground station, where the yardmaster had organized a sterling defence of his domain. The other survivor is the Elephant with the castle on its back. Once it stood above the pub that bore its name; that closed down in March 1959 when redevelopment began on the Elephant & Castle. Now it guards the entrance to the shopping centre.

The London County Council had such grand plans for the area when the bulldozers moved in nearly half a century ago. They designated 30 acres around the Elephant as a 'Comprehensive

Development Area'. Traffic jams would be a thing of the past, so they said, and a web of pedestrian subways would make life safer. The island between Newington Butts and Walworth Road, home to the pub and its Elephant and the Drapers' Almshouses, was removed so the road could be widened. In 1965 the new shopping centre was unveiled. It was a giant pink carbuncle with which even the Elephant outside looked embarrassed to be associated. The Trocadero was demolished during the redevelopment and the Manor Place Baths closed in 1978.

Across post-war London the vast number of homes destroyed during the Blitz led to a radical regeneration, or 'modernization' as it was called by the politicians. In Southwark, like so many working-class swathes of London, the housing shortage of the 1950s was solved by constructing cheap and easy-to-build tower blocks. Out of the ashes of Georgian and Victorian streets rose enormous twenty-five-storey high rises. Hundreds of families moved in on top of one another and the old terraced houses of Southwark, where mothers chatted on the doorstep while their children played in the street, were consigned for the most part to memory. In the new tower blocks neighbours were nameless people occasionally glimpsed in the lift. Gurney Street, for example, where the bomb exploded in June 1942, is no more. In its place is a sprawling, dark, dangerous-looking housing estate.

It took the dawn of the twenty-first century – and the Elephant & Castle's reputation as the fifth most dangerous road junction in London and one of its most crime-infested areas – for the authorities to finally admit something had to be done. Southwark Council announced a £1.5 billion scheme to regenerate the area once more and work is scheduled to start in 2006 with an estimated completion date of 2014. Might the Elephant become the deserved king of London's concrete jungle?

THERE are very few people alive who can remember what happened to London on the night of 10 May 1941. Some of those who figured in this book were dead within a year or two. Squadron Leader Arthur 'Scruffy' Sanders transferred from 264 Squadron to command 85 Squadron in June 1941. In October he wangled Flying Officer Freddie Sutton a posting to join him flying Havocs. Two days before Sutton's arrival Sanders and his gunner were shot down in a duel with a Junkers 88 over the North Sea. Sanders was seen to bale out but no trace of him was found. He was 28.

Sutton remained with 264 Squadron until 1942 when he was sent to Canada as a gunnery instructor. He survived the war and died in 1981.

Josef Dygryn, the Czech pilot who was credited with three confirmed kills in the early hours of 11 May, shot down four more German aircraft by the end of June. In September 1941, now with a Distinguished Flying Medal, he was posted to No. 310 Czechoslovak Fighter Squadron. He returned to No. 1 Squadron in May 1942 and took part in a number of 'Intruder' sorties over northern France. As he crossed the French coast at Le Havre after a raid on Evreux-Fauville airfield on 4 June 1942 his Hurricane was hit by German AA fire. Dygryn struggled valiantly to nurse his aircraft over the Channel but he crashed a few miles off the English coast. His body was washed ashore at Worthing in September that year and he was buried in Kent.

Wladyslaw Nowak, the first Polish day fighter to claim a night victim when he shot down an He111 over the Channel, had risen to Squadron Leader at the war's conclusion. He was reunited with his wife Henrietta, a concentration camp survivor, after six years separation, and the couple settled in Worksop in Nottinghamshire where they enjoyed an unobtrusive life running a delicatessen. Nowak died in 1982 aged 74.

Guy Gibson and Richard James spent the days after 10 May searching for the reason for the impotency of their cannons. On both 11 and 12 May, and twice on the 13th, they took to the air in an attempt to solve the problem. It wasn't until Thursday the 15th that the cannons eventually fired. The fault, recalled Gibson, was 'a simple electrical solenoid in the firing button becoming unserviceable. Sergeant James ... was furious'. The pair had more luck on 6 July when they shot down a Heinkel He111 over Sheerness. They damaged a Junkers 88 a couple of months later. The summer of 1941 was, Gibson said later, 'the happiest I have ever had. But somewhere in the back of my head was that little bug of restlessness'. In early 1942 he was posted to command 106 Squadron flying Lancaster bombers, and in May the squadron took part in the first RAF 1,000-bomber raid on Germany. The following year he was awarded the Victoria Cross for leading 617 Squadron on the legendary 'Dambuster' raid on the dams at Moehne and Eder. The citation concluded by describing how: 'Wing Commander Gibson personally made the initial attack on the Moehne dam. Descending to within a few feet of the water, he delivered his attack with great accuracy. He then circled very low for thirty minutes, drawing the enemy fire and permitting as free a run as possible to the following aircraft. He repeated these tactics in the attack on the Eder dam. Throughout his operational career, prolonged exceptionally at his own request, he has shown leadership, determination and valour of the highest order.'

On the evening of 19 September 1944 Gibson acted as Master Bomber on a raid over western Germany. On the return home his Mosquito crashed and exploded in a farmer's field near Steenbergen in Holland. Gibson was 26; and the holder of a VC, DSO & bar, DFC & bar and US Legion of Merit.

Richard James was posted to 96 Squadron in the spring of 1942 and continued operational flying for another year. He finished the war with a DFM and a commission. James returned to his family and his pre-war job in the building industry.

'During the war I never allowed myself to dwell on death,' James reminisces. 'An aunt of mine asked me once when I was flying "How do you feel about killing people?". I was damned annoyed with her for asking that. I was losing pals myself all the time. But since the war I've thought about it a lot. When you got a "blazer" and everything burned from the word go ... well, I realise now that there were blokes inside that Heinkel being roasted to death.'

James is 93 and lives in Surrey. Last year he arranged for a friend to travel to Holland to lay a wreath on the grave of Guy Gibson on the sixtieth anniversary of his death. 'I couldn't have been treated with more respect by Guy,' he says. 'He was a first-class man and a first-class pilot. Some people have said in the past that he was fearless. He wasn't, I saw him frightened. He would have been scared during the Dambusters Raid but he would have overcome his fear.'

Johnny Freeborn was transferred from 74 Squadron to an Operation Training Unit in the north-west of England on 6 June 1941. He was the last of the pre-war originals to leave the 'Tiger' Squadron, the end of a momentous chapter in their history. It was with sadness that Freeborn waved goodbye to his comrades at Gravesend and the lake at Cobham Hall where Roger Boulding's German dinghy was proving popular.

On 16 June, ten days after Freeborn's departure, 74 Squadron went on a fighter sweep over France. 'I was leading one section of four,' recalled Boulding, 'and [we] spotted a formation of Me109s with their famous yellow painted noses climbing towards us.' Boulding and his number two flew towards the sun in a weaving pattern. 'I looked behind, saw what I thought was the same aircraft guarding my rear ... when my aircraft was hit from behind. The armour plate behind my seat took it and saved me.' At 25,000ft Boulding's Spitfire went into a spiral dive. 'I pulled the canopy release without too much trouble, undid my seat straps but could not get out because of the spinning, so had to get my knee up and jerk the stick forward, which effectively catapulted me out ... the Germans had ample time to reach me when I landed and before I could stand up there were plenty of them threatening me with an assortment of weapons.' Flying Officer Roger Boulding was released from captivity in 1945 and remained in the RAF after the war. He retired with the rank of wing commander and died in 2001.

The rest of Johnny Freeborn's war lacked neither excitement nor enjoyment. In 1942 he was sent to the USA as a test pilot. For the next year American manufacturers sought his input on a range of aircraft, from the P47 Thunderbolt to the Mustang to the B17. On his return to the UK Freeborn had postings with 602 and 118 Squadrons before he was promoted to Wing Commander Flying of 286 Wing in Italy.

In 1946 Freeborn left the RAF because in peacetime it was run 'by nincompoops'. He had a brief stint as a vehicle inspector for the Ministry of Transport but later became regional manager for a soft drinks firm. He lives in the north-west of England, one of the last remaining Battle of Britain 'aces'. He's not a spiritual man, nor is he philosophical. He owed his survival to 'practice ... that's what made a good Spitfire pilot. Practice. That and the fact I was bloody good anyway!'.

Albert Hufenreuter, the navigator shot down by Roger Boulding, was sent to northern England while his wounds healed. Later he was shipped to Canada where he learned English in a POW camp. 'My father was treated so well in captivity he couldn't really believe his good fortune,' says his son, Albert junior. Upon his release Hufenreuter knew he couldn't return to the family home in Quedlinburg, then part of the newly formed communist East Germany. Instead he settled in Hamburg where he went to university and later became an

English teacher. He returned often to Britain and in the early 1980s visited the Kent field in which his Heinkel had crashed forty years earlier. 'There was a knock on my front door one day,' recalls Peter Huckstepp who, as a teenager, had been the first villager from Kennington to reach the downed aircraft. 'A big man was standing there and he introduced himself as Albert, the navigator of the plane that had crashed in 1941. My wife and I invited him in, gave him a cup of tea, and he told us the whole story. He was very happy with the way things turned out; the raid on London was his squadron's last mission before their posting to Russia and not many of them survived that.' Albert Hufenreuter finished his teaching career as headmaster of a Hamburg school and in retirement worked as a tour guide in the city for fifteen years. 'He went to Luftwaffe reunions every year,' says his son, 'but he didn't often talk about his experiences as a pilot. Like most bomber pilots he didn't enjoy dropping bombs on cities, but it was war.' Albert Hufenreuter died in February 2002 aged 85.

Hugo Sperrle, the architect of the raid, was in charge of the Luftwaffe when the Allies landed in Normandy on 6 June 1944. His forces were overwhelmed by the superior numbers of American and British aircraft and Sperrle was dismissed from his post in August. He was tried at Nuremburg after the war but acquitted of war crimes in 1948. He died in Munich in 1953 aged 68.

Not long after the raid the RAF revised their figure for the tally of German bombers shot down from thirty-three to eleven. The newspapers didn't run with the news. In fact Luftwaffe records examined after the war stated that fourteen of their aircraft failed to return from the assault on London, the highest number in any single raid, but still only 3 per cent of the total bomber force.

The same Luftwaffe intelligence records revealed the enormousness of the raid. A total of 507 aircraft had unfurled 711 tonnes of high explosives over London for six and a half hours. The most prolific bomb was the 50kg, of which 1,492 were dropped. Seventy-seven parachute mines fell, eight were 500kg and the rest 1,000kg. Seven 'Satans' landed. Packed with 1,800kg of high explosive, never was a name more apt. The number of 1kg incendiaries surprised no one who had endured 10 May. Was it only 86,173? they asked each other with a grim smile.

THE numbers of incendiaries dropped stretched the London Fire Service to breaking point. Every one of their 1,270 fire engines, water units and firefloats within the London County Council boundary was in action.* Nearly 400 pumps from outer London and seventy from other counties were called in to help tackle the 2,154 fires, nine of which were 'Conflagrations', fires officially classified by fire service chiefs as out of control. Another twenty were 'Major Fires' requiring the attention of at least thirty pumps if they weren't to be added to the Conflagration list. The blaze in the vaults under Waterloo Station was the last one to be conquered by the firemen, on 22 May.

The cost to the fire service was heavy. Thirty-six firemen dead and 289 wounded. Nine months previously some of them had been refused entry to restaurants because they were 'parasites' and 'army dodgers'. Their sacrifice wasn't in vain. Not only was London saved but the experiences of 10 May was the final impetus for nationalization of Britain's fire brigades. The clamour for the country's 1,600 independent brigades to be formed into one homogenous service with one command structure and universal training methods had been building for months. London firemen in particular, who had been sent to reinforce colleagues in other English cities, had been staggered by their primitive equipment and techniques. Conversely, when the provincial

*The firemen were scathing about the performance of the city's firewatchers, criticism that led to much stricter enforcement of the Fire Precaution and Business Premises Order.

fire crews arrived to reinforce London, their bravery was indubitable; their efficiency questionable.

Just three days after the raid the Fire Services (Emergency Provisions) Bill was passed in the Commons' new home in Church House. The whiff of charred timber from the smouldering Chamber across Dean's Yard helped guide its passage through Parliament. The crux of the bill was the amalgamation of Britain's 1,600 brigades into twelve regions; the cost was met by the Government and thus the organization of individual brigades was no longer in the hands of (often) incompetent local authorities. On 18 August 1941 the National Fire Service was born.

In 1947 a new bill was passed that handed control back to local authorities, though instead of 1,600 brigades there were 147. The largest was the London Fire Brigade and Fred Cockett was one of its officers. Later in the war he had been transferred to Fire HQ on the Albert Embankment and he decided to remain in the service. He retired in 1969 and took a job with a bank for twelve years. He is now 91 and lives in Kent.

William 'Mac' Young left the fire service at the end of the war and returned to the hotel industry. He left after a short while, unable to come to grips with his pre-war life. Instead he worked for himself as a boat-builder. His first wife died in the 1970s and he enjoyed an equally happy marriage with his second wife before her death a few years ago. He is 93 and lives alone in a quiet corner of Suffolk.

Lew White never recovered fully from the injuries he received on 29 December 1940. 'I came back to the sub-station but I was no good to them. I was vague in a lot of things I did. Not long after 10 May I was transferred to Abingdon in Oxford as a fire prevention officer, going round looking at premises to make sure they were in order. Eventually I was invalided out of the service because I was suffering from arthritis as a result of my injuries.' Lew returned to Smithfield Market after the war and then set up his own meat business. He is 91, lives in north London and still enjoys watching Arsenal on the television.

London's firewomen were the unsung heroines of the Blitz. To the public they were invisible, tucked away in the control rooms for the most part, but without them the fire service would have collapsed. 'We had enormous admiration for the auxiliary firewomen,' says Fred Cockett.

Florrie Jenkins and her mum were rehoused in Hornchurch after being bombed out of their house in Coutts Road. They were each given £20 to cover what they had lost. The extent of the damage to Coutts Road was such that what little remained was pulled down and it ceased to exist. Florrie was later transferred to Ilford in Essex and she left the service at the end of the war. She is 83 and lives in Norfolk.

Reenie Carter made the fire service her career. She finished the war at Lambeth and spent a further three years in the London Fire Brigade before a move to the Surrey force. In 1965 she transferred back to London and retired in 1980 after forty years and six months as a firewoman. She is 84 and lives in Westminster, a brisk ten-minute walk from the Houses of Parliament.

Bobbie Tanner was transferred to the Avery Hill Transport Depot in Woolwich once the Blitz had ended. She remained in the fire service until February 1945 when she left to marry an army officer. No other firewoman was awarded a George Medal during the Second World War. Post-war life for Bobbie was neither colourless nor conventional. After the breakdown of her marriage she devoted much of her time to hunting and winning prizes in point to point tournaments. Then in the 1950s she took up rally driving ('Driving the lorry in the Blitz was wonderful preparation,' she says) and completed the Monte Carlo rally four times. She mellowed somewhat in middle age and became a delivery driver and then a carer. She is 86 and lives in mid Wales. 'The Blitz was terribly exciting,' she says, 'and a great experience. I had no animosity towards the Germans; the pilots were doing their job and ours were doing theirs.'

Ballard Berkeley's friend George Niblett was one of seventeen policemen killed during the raid on 10 May. A further seventy-five were injured. 'Of course there were scary moments,' reflected Ballard, 'but you knew you had to do a job. The country had to go through this period; you knew it was going to be awful, you accepted it was going to be awful and you knew you were going to bloody well win. There was never any question: we were going to win.' In 1942 Ballard resumed his acting career, starring in the classic war film *In Which We Serve*. When he left the police he went back into the theatre, appearing in New York and London. More television and film work followed, as well as a part in 'The Archers' before, aged 71, Ballard was offered a part that propelled him almost overnight into the pantheon of British comedy legends. The series was Fawlty Towers and Ballard played Major Gowen, a doddery old sot who loved cricket and hated Germans. And rats. Ballard Berkeley died in 1988, aged 83.

Auxiliary ambulance driver Anne Spooner remained in the service until the end of the war. While her husband transferred his flying skills from military to commercial aircraft, Anne raised their children. In later years she typed up the manuscripts of the seven aviation books he wrote. She is 87 and lives in Sussex.

For Dr Kenneth Sinclair-Loutit the 'Blitz became an institution'. For nine months he and his Heavy Rescue squad were in the eye of the storm. 'It killed people, it frightened people, it made life much more difficult but the one thing it did not do in London was to lower that strange collective reaction called morale.' In the beguiling lull that swept through London after 10 May Sinclair-Loutit was transferred to the London Civil Defence Region HQ as secretary to the standing committee linking the ninety-odd municipalities of the capital. He left the medical world after the war to run a pharmaceutical company in Morocco. On the sixtieth anniversary of the raid he wrote to the Mayor of Islington reflecting on the night, the Blitz in general and his admiration for the steadfastness of the borough's people. His letter received no acknowledgement. He died in Morocco in October 2004, aged 91.

Bill Church looks back on his days as an anti-aircraft gunner with fond amusement. 'It was a Heath Robinson outfit,' he says. 'How we won the war I don't know, more by luck than judgement.' After the Blitz he was posted to the Honourable Artillery Company and took part in the invasion of France in 1944 and the push into Germany the following year. Bill was demobbed in 1946 and became a plant engineer. He is 84 and lives in his native Essex.

After Harry Beckingham's bomb had been defused at Victoria Station it was taken to a 'bomb cemetery' in Richmond Park. 'We had equipment there which we used to steam out the explosives. We reused the empty cases.' Harry was commissioned in 1943 and remained in the UK throughout the war, defusing bombs unearthed from the Blitz of 1940/41. He spent several months in Guernsey after its liberation in 1945, making safe the thousands of mines laid by the Germans. 'What I remember most about working in London was the people,' he reflects. 'As soon as we pulled up in our lorry the tea would be out in jugs and when we finished defusing the bomb they had a whip-round and gave us a few quid saying "this is what we've collected for you". They were great people.' He studied Structural Engineering at Manchester University after his demob, went to Rome on a scholarship and returned to work as a divisional engineer in the British construction industry. 'I never missed the buzz of bomb disposing after the war,' he says. 'I found my civilian work very interesting.' In Harry's opinion, 'there must be dozens of UXBs from the Blitz still out there, particularly at the bottom of the Thames.' (In 2001 a dredger picked up a UXB during refurbishment to Hungerford Bridge that led to the temporary closure of the London Eye.) He is 85 and lives in the north-west of England.

The violence of 10 May extinguished something in Cosmo Lang, a gentle man who abhorred war. He resigned as Archbishop of Canterbury at the end of 1941 and left office in March the next year. He lived out the rest of the war away from the bombs in Surrey. On the evening of 4 December 1945 he and a friend discussed what might happen to the world now that peace had arrived. Lang was uncharacteristically morose. 'I think I've had enough of this one,' he told his friend. The next morning his 81-year-old heart gave out. He was buried in Westminster Abbey.

In 1946 Alan Don was appointed Dean of Westminster following the death of Paul de Labillière. It was a short move for him and his wife from 20 Dean's Yard to the Deanery. His tenure lasted thirteen years until his retirement in 1959. He died in 1966 aged 81. Dr Jocelyn Perkins resigned from his post as Westminster Abbey Sacrist in 1959, sixty years after his appointment. He died three years later aged 92, and his remains were interred in the Abbey's St Faith's Chapel.

After the destruction of St Mary's, Christopher Veazey was offered the parish of St Silas in Peckham. The previous vicar, Stanley Tolley, had been a victim of 10 May.

He was inducted into his new church in August 1941 and four months later Joan gave birth to their first child, a daughter called Janet. A son followed in 1944. 'The Blitz was a very exciting time,' says Joan. 'I think too we felt a certain amount of pride because we were doing something for our country.'

The thrill of the Blitz was exchanged post-war for the more sedate surrounds of rural Kent. For over thirty years Christopher Veazey was vicar in the parish of Doddington. 'The Veazeys are well remembered,' says one of the former church wardens. 'He was a wonderful vicar.' Christopher died at the end of 2003 at the age of 91. 'We didn't have one quarrel in 63 years of marriage,' says Joan.

She is 91 and lives alone in Kent. Tucked away in a box in the corner of one room are several pieces of shrapnel she collected on the night of 10 May 1941.

Gladys Shaw worked for the Ranyard Mission in south London throughout the war. 'There was a very happy fellowship among the people,' she remembers, 'I think that's what got us through.' In 1946 she applied and was accepted for overseas mission work. She sailed to India and spent thirty years as a missionary in the poorest rural regions of the country, teaching the children to read and write. Gladys retired in 1976 and returned to England to live with her sister in Sussex. She went back to India on an extended holiday in 2001, aged 90, and was fêted by the village to which she had given so much. She is 94 and lives in Sussex.

WALTER Elliot, MP for Kelvingrove, Glasgow, the man who smashed his way into Westminster Hall, lost his seat in the 1945 Election and died in 1958. Robin Duff, the BBC correspondent who witnessed first-hand the destruction of the House of Commons Chamber, was the *Daily Express* bureau chief in Paris and then Delhi for a time after the war. In the 1950s he returned to his native Scotland and opened a hotel near Aberdeen. He was president of the Scottish Ballet for a number of years and died in 1990 aged 75.

Peter Huckstepp, who with his dad had rescued Albert Hufenreuter and his crew, left the Kennington Home Guard in 1942 to join the Royal Navy. He still lives in the village in which he was born but 'we've been swallowed up by Ashford in recent years'. Houses stand now on the field where Richard Furthmann nursed down his sick plane with consummate skill. Jean Ratcliffe never saw her boyfriend Clement Edwards again after emerging from the Waterloo Station lavatories. He joined the merchant navy as a radio officer and in 1942 went down with his ship in the Atlantic after a U-boat attack. He was 22. Jean married after the war and has lived in Wales for over fifty years. 'The funny thing was that my future husband was also in the gents' loo at Waterloo on 10 May 1941! He had been en route somewhere and got caught in the raid and like everyone else at Waterloo he took cover in the lavatories. The next morning there were no trains so he took a taxi that cost him an arm and a leg.'

Vera Reid married after the war and moved to the USA where she wrote short stories for several publications including the *New Science Monitor*. She died in the 1980s. John Hughes swapped Fleet Street for the healthier climate of Australia and died in the 1970s. His daughter donated his letters to the Imperial War Museum in London. Madeleine Henrey and her husband flitted between London and their farmhouse in Normandy for many years after the war. Their son, Bobby, had considerable success as a child actor. Between 1941 and 1979 Madeleine had more than thirty books published, *The Little Madeleine* being the best known. Her husband died in 1982 and she spent her widowhood in Normandy reading and keeping hens. She died in 2004, aged 97. Quentin Reynolds continued to write for *Colliers Weekly Magazine* for a number of years after the war. He also had published a number of bestselling books, foremost among them were his wartime memoirs, *Only the Stars are Neutral*. He died of cancer in San Francisco in 1965 aged 62.

When Austin Osman Spare was pulled from the rubble of his flat he was paralyzed down his right side and bereft of dozens of his drawings. It was six months before he regained feeling in his hand and fingers and another couple of years before he started painting. His fecundity in the post-war years, however, was remarkable, and in 1947 he exhibited two hundred works at an exhibition at the Archer Gallery. For the next few years Spare's pastels were lauded as much as his obsession with sex magic was condemned. When he died in 1956 aged 68 he was eulogized as one of England's most creative and eccentric geniuses, though his sexual deviancy meant he was never accepted by mainstream society. An exhibition of his work was held in London in 1999, again to critical acclaim.

In 1951 Cecil King usurped Harry Bartholomew as chairman of the *Daily Mirror* newspapers and created a media empire that in 1963 became the International Printing Corporation (IPC). He was ousted in 1968 and died in 1987, aged 86.

Tom Finney's Preston North End won the cup final replay against Arsenal 2–1. 'There wasn't enough gold to spare for winners' medals,' recalls Tom, 'so we each got four 12 shilling wartime savings certificates. Fortunately Preston cast some bronze medals with a little gold cup in the middle. I've still got mine and it's one of my most treasured possessions.' In 1942, Tom joined the Queen's Royal Lancers and served in North Africa and Italy. After the war he became in the eyes of many the greatest ever English footballer. He scored thirty goals for his country in seventy-six matches and was knighted for his services to football. He stayed loyal to Preston throughout his career, during which time he also owned his own plumbing business, and retired from the game in 1960. He still watches every home game and is idolized in his home town.

Vera Lynn's achieved everlasting fame through her wartime radio request programme 'Sincerely Yours'. Her songs 'We'll Meet Again' and 'White Cliffs of Dover' are enduring classics. She starred in several wartimes revues and films and was the forces' pin-up. After the war her success grew and in 1952 she became the first British artist to have a No. 1 record in the USA with 'Auf Wiedersehen Sweetheart'. Her first and only UK No. 1 was the 1954 song 'My Son, My Son'. She conquered television in the 1960s and was created a Dame of the British Empire in 1975. She is 87 and continues to attend forces' reunions across the UK.

FOR the ordinary working man and woman in London 10 May 1941 was the most devastating raid of the war. Of the sixty-one London boroughs that reported bomb incidents on 10 May, 137 people in Westminster were killed, in Bermondsey sixty-seven, in Stepney 12.7 tonnes of high explosives were dropped per square mile of the borough. In Southwark 14.4 tonnes fell per square mile, leaving 135 dead. And yet, in spite of the terror, there was a purity in London during the Blitz that hadn't been there before nor has it been since. It was a camaraderie of classes. 'The danger and destruction deprived us of all social camouflage,' reflects Dr Kenneth Sinclair-Loutit. 'Social boundaries did no one any good during the Blitz so they ceased to exist. Sometimes we were all frightened together and there was no concealment of the fact. Rich and poor, working class and the upper class suffering together.' For nine months Londoners came together. It wasn't for long but it was good while it lasted.

Joan Wallstab returned to London with her husband Eddie to find her parents' flat in the Sessions House blown to bits. She had to go to the mortuary to identify her father's body, an experience that still haunts her. Her mother survived and received £2 10s 0d a week for ten weeks, the standard pension for the spouses of Civil Defence workers killed in the line of duty. Joan and her mother moved into a flat in Kennington for the rest of the war while Eddie saw out his army service. After the war she worked for an insurance company and her husband enjoyed a long association with the BBC as a film editor. She is 84, lives in Kent and is still great friends with Emily Macfarlane.

Joe Richardson joined the Royal Navy in 1944 and served on a destroyer in the Pacific. After the war he travelled around the world for a few months, curious to taste other cultures. He toured Europe and in Germany fell in love with a young woman. They married and returned to London, where Joe worked as a taxi driver for many years.

Joe's son lives in Germany, but he and his wife live in Southwark, just a couple of hundred yards from Vestry Road. It was to a house in that road that John Fowler escorted his cousin Rose after watching *Seven Sinners* at the Peckham Odeon. The landmine that killed her dad left Rose trapped in the basement for two days. 'She seemed fine after they brought her out,' says John, 'but then about six months later she collapsed at work from post-traumatic stress.' Rose recovered and married after the war. She died in 2001, aged 76.

John and his parents eventually made it to Sturminster Newton for his sister Joan's thirteenth birthday party, after which he went back to his farm in Haslemere. The cessation of the Blitz encouraged many parents to bring their children home in 1942 but while Joan returned, John was having too much of a good time in the country. He liked visiting London, however, and was back in Peckham in mid May 1943 for Joan's fifteenth birthday. Two days after he had gone back to the farm a lone German bomber launched a hit and run raid on a Peckham factory which manufactured ammunition. The bombs missed and destroyed John's house in East Surrey Grove. His parents and sister died without knowing what had hit them. 'For a long time afterwards I would've killed any German I met,' says John, 'but those feelings faded in time. I don't bear any resentment now. It was just one of those things that happen in war.' John never left his farm, and it wasn't until 1997 that he revisited his old haunts in south London. He is 78 and now lives in Sussex.

Gladys Jenner married her Royal Engineers boyfriend, Tony, at the end of 1941, but then saw little of him until the end of the war. 'The Peek Freans company had two rules,' remembers Gladys. 'They didn't employ married women or trade unionists. So my getting married presented them with a problem. I was called into the manager's office and told that because so many men had been called up I would be allowed to remain as a clerk, but my salary would be reduced from

£2 10s to £2 5s. I left and got a job with the National Cash Register Company.' After the war Gladys and her husband moved out of London and raised a family. She is 82 and lives in Sussex.

Tom Winter 'counts himself lucky to have lived through a time in history when people on the whole were good, decent and much more socially conscious of their fellow man'. The Blitz might have destroyed thousands of homes and properties but, says Tom, the bombing gave Londoners a 'spiritual wealth'. But there is one thing that still troubles him, a worry bequeathed him by his dad: 'The bomb that killed Granny Humphreys and her rescuers didn't explode; it tunnelled under the foundations and caused the avalanche of earth and debris.' Over the years Tom has expressed his concerns in writing to the relevant authorities but as far as he knows no investigation was conducted. 'The bomb's deep down and dormant,' says Tom, 'but still a potential danger.' Tom lost touch with Ken Humphreys after the war but his other great friend from his Bermondsey boyhood, Teddy Turner, died in 1992. Tom worked in road haulage for many years but also won recognition as a talented amateur sculptor and artist. His bust of John F. Kennedy was commissioned by Bermondsey Council and was on display for a long time until it was stolen. He is 76 and lives in Buckinghamshire.

Emily Macfarlane wheedled her way out of the fire service later in 1941 and got a job working for the war office in Whitehall. There she remained for twenty-nine years before becoming a hospital registrar. Life has often been cruel to Emily; her fiancée died during the war just days before he was due home on leave and the Blitz robbed her of several close friends. She's 85 now but old age has yet to rob her of her grace and elegance. She lives alone on the same Peabody Estate in which she was born and she still does her shopping in the Elephant & Castle.

'There's been some sad times,' she says, 'but you don't think about them. You think of the good times and of the happy memories, not the bad ones.'

Emily is one of a dwindling band of Londoners able to brush off the mothballs of memory and talk about the Blitz. They've journeyed a long way since those stomach-churning days, but the Blitz has been their constant companion all the way. Yet for the survivors there is no regimental association, no squadron reunion at which to share their experiences over a pint and a pie. They're left alone with their mothballs. To successive generations the Blitz has become a cliché, like the word itself, now part of everyday vocabulary. Who cares now if it was 'business as usual' in 1941? London, to use another cliché, has moved on. It's a fast dynamic modern city, cosmopolitan and multi-cultural. The Blitz has no relevance in modern London, has it?

Without Emily and Tom and Gladys and Joe and thousands of other anonymous but steadfast Londoners, now lined and ancient and forgotten, London would not have survived the Blitz. For nine wicked months the city was brutalized. In seventy-one major raids (the Luftwaffe's definition of a major raid was one in which 100 tonnes or more of HE were dropped successfully), 18,291 tonnes of high explosive killed over 19,826 civilians and wounded 72,570. If London had yielded, what then for Europe? But the city stood firm as the world looked on, even on the night of 10 May 1941, London's bloodiest and longest night. Perhaps, too, it was also London's greatest night, the culmination of a nine-month battle against fascism that ended with the symbol of the free world, bruised and battered, but unbeaten and as bloody-minded as ever.

The Main Players

Harry Beckingham A 20-year-old bomb disposal expert from Manchester

Ballard Berkeley Actor-turned-policeman whose beat was London's West End

Reenie Carter A 20-year-old firewoman based at Westminster Abbey fire station

Bill Church A pessimistic anti-aircraft gunner at Wormwood Scrubs

Fred Cockett A fireman based at Waterloo Station

Canon Alan Don The rector of St Margaret's Church, Westminster

Tom Finney A 19-year-old footballer playing in a Wembley Cup Final

John Fowler A 14-year-old evacuee back in London for his sister's birthday

Johnny Freeborn A Battle of Britain Spitfire ace with a killer instinct

Albert Hufenreuter German navigator commanding a Heinkel of 55 Group

John Hughes Cynical Australian journalist working for the PA in Fleet Street

Richard James A radar operator in a Beaufighter of 29 squadron piloted by Guy Gibson

Gladys Jenner A fun-loving young woman working for Peek Freans in Bermondsey

Vera Lynn Britain's singing sweetheart starring at the London Palladium

Jamie MacDonald A *New York Times* reporter with the US press pack at the Savoy

Emily Macfarlane A reluctant firewoman in the Elephant & Castle

Joe Richardson A streetwise 16-year-old from south London

Larry Rue A *Chicago Tribune* reporter with the US press pack at the Savoy

Kenneth Sinclair-Loutit Socialist doctor in charge of a Heavy Rescue team in the City

Gladys Shaw A missionary in south London who held prayer meetings in shelters

Freddie Sutton A gunner in a two-seater Defiant of 264 Squadron

Bobbie Tanner A flighty society girl who won a George Medal as a firewoman

Joan Veazey A pregnant vicar's wife living at St Mary's Church in Kennington

Tom Winter A 12-year-old from Bermondsey who collected bombs as souvenirs

Mac Young A surly Scottish fire engine driver based in Paddington

Glossary

AA Anti-Aircraft

Ack-Ack Abbreviation of anti-aircraft

AFS The Auxiliary Fire Service, into which the first volunteers enrolled in early 1938 as part of the Government's Air Raid Precautions Act of 1937.

ARP Air Raid Precautions, a voluntary scheme first introduced in 1935 to encourage local authorities to organize Civil Defence units. Two years later an Air Raid Wardens' Service became compulsory.

AI Airborne Interception was the rudimentary system in use by RAF Beaufighter nightfighters in 1940–41, which later came to be known as radar.

Beleuchter-gruppe Fire-lighters in English, the nickname given to KGr100 of the Luftwaffe.

CO Commanding Officer

DFC Distinguished Flying Cross, awarded to officers and warrant officers of the RAF for acts of valour or devotion to duty while on active operations against the enemy.

DFM Distinguished Flying Medal, awarded to non-commissioned officers and men of the RAF for acts of valour or devotion to duty while on active operations against the enemy.

Elephant & Castle The origins of how the junction got its name are still disputed today. One version is that in the seventeenth century an inn on the site had a sign depicting an elephant with a castle on its back, that was actually a howdah. The other tale is that around the same time a Spanish princess came to London to marry an English lord. On her way into the city she passed through the junction and in honour of the eldest daughter of the monarch [*Infanta de Castile*] the locals named it after the princess.

EWS Emergency Water Supply. These were mobile dams used by the Fire Brigade.

GCI Ground Controlled Interception stations were the radar stations that directed RAF nightfighters within a mile of a hostile aircraft before the nightfighter's radar operator took over the interception using the AI.

HE High Explosive

HQ Headquarters

IAZ Inner Artillery Zone was the airspace over a city that was covered by anti-aircraft defences.

IFF Identification Friend or Foe transponders were fitted on RAF aircraft to enable British radar stations to distinguish friendly aircraft from hostile ones.

IHQ International Headquarters of the Salvation Army in Queen Victoria Street

Kampfgeschwader (KG) Bomber Group of the Luftwaffe

Kampfgruppe (KGr) Bomber Wing of the Luftwaffe

Knickebein 'Bent Leg' in English, it was a radio system used by the

German bombers to guide them to targets in Britain using beams from radio stations on the French coast.

Kopfring A ring fastened to the nose of German thin-casing high explosive bombs to stop them boring too deep into the ground.

LAAS London Auxiliary Ambulance Service

OTU Operational Training Unit (in the RAF)

RAF Royal Air Force

Sprengbombe-Cylindrisch (SC) Thin-cased German high explosive bomb

Sprengbombe-Dickwandig (SD) Thick-cased German high explosive bomb

VC Victoria Cross, Britain's highest medal awarded for acts of valour in the face of the enemy.

WAAF Women's Auxiliary Air Force

W/O Wireless operator

WVS Women's Voluntary Service, formed in 1938 primarily to stimulate the enlistment of women into the Air Raid Precaution (ARP).

X-Gerat A later and more sophisticated German radio system than *Knickebein* which used four beams to guide their bombers to the targets.

Index